PETROLOGY OF SEDIMENTARY ROCKS

ROBERT L. FOLK

Hemphill Publishing Company
Austin, Texas 78703

Printed in the United States of America

Library of Congress Catalog Card Number 80-83557
ISBN Number 0-914696-14-9

TABLE OF CONTENTS

PREFACE

This syllabus is by no means intended as a textbook on sediments. Rather, it was originally intended to supplement lecture and laboratory material given in sedimentary petrology courses at The University of Texas. Consequently it is to be used in conjunction with standard textbooks in the field such as Pettijohn, Sedimentary Rocks, Krumbein and Pettijohn, Manual of Sedimentary Petrology, Blatt, Middleton and Murray, Origin of Sedimentary Rocks, Pettijohn, Potter and Siever, Sand and Sandstones, Bathurst, Carbonate Rocks, Carver, Procedure in Sedimentary Petrology, or Royse, Sediment Analysis, for this reason no references to the literature are given, as these references are readily available in those texts. Persons responsible for particular ideas are indicated by parentheses. Figure revisions are by Connie Warren.

None of the statements herein are to be regarded as final; many ideas held valid as recently as two years ago are now known to be false. Such is the penalty of research. This syllabus merely states the present condition of the subject. The rapid rate at which sedimentary petrologic data is now accumulating is bound to change radically many of the ideas contained within.

Much of this syllabus is based on material obtained in sedimentation courses taught at the Pennsylvania State College by Paul D. Krynine and J. C. Griffiths, 1945-1950, together with later modifications of and additions to this material by the present author during his own work on sediments after 1950. I would therefore like to dedicate this booklet to those two inspiring teachers: Krynine, without peer as a sedimentary petrographer, mineralogist and man of ideas (see JSP Dec. 1966); and Griffiths, pioneer in the application of rigid statistical techniques to description of the physical properties of sediments.

Historically, Henry Clifton Sorby of Sheffield, England (1826-1908) is the founder of sedimentary petrography (and microscopic petrography in general). His work was so voluminous and so excellent that it was not matched until well into the twentieth century, fifty years after his publications. Although the microscope had been used earlier to study slides of fossils and a few rocks, Sorby was the first geologist to realize their importance, cut his first thin section in 1849 (a cherty limestone) and published on it in 1851, the first paper in petrography. Sorby demonstrated his technique to Zirkel in 1861, and thus igneous petrography was born. Sorby's three monumental papers were on the origin and temperature of formation of crystals as shown by their inclusions, etc. (1858); on the structure and origin of limestones and petrography of fossil invertebrates (1879); and on quartz types, abrasion, surface features, and petrography of schists and slates (1880). He made 20,000 paleocurrent measurements for a decade before his publication (1859). He also has fundamental publications in structural petrology (1856), studied fluvial hydraulics, founded the science of metallography in 1864, and devoted the latter part of his life to study of recent sediments and marine biology. A good biography is given by Judd (1908, Geol. Mag.), and Naturalist (1906) lists some 250 of Sorby's papers; a short review of Sorby's career is given in J. Geol. Educ. 1965. Even today his papers deserve detailed study by every petrographer. Two volumes of his collected works have been edited by C.H. Summerson (publ. by University of Miami).

INTRODUCTION TO SEDIMENTARY ROCKS

Sedimentary rocks cover some 80 percent of the earth's crust. All our knowledge of stratigraphy, and the bulk of our knowledge of structural geology are based on studies of sedimentary rocks. An overwhelming percentage of the world's economic mineral deposits, in monetary value, come from sedimentary rocks: oil, natural gas, coal, salt, sulfur, potash, gypsum, limestone, phosphate, uranium, iron, manganese, not to mention such prosaic things as construction sand, building stone, cement rock, or ceramic clays. Studies of the composition and properties of sedimentary rocks are vital in interpreting stratigraphy: it is the job of the sedimentary petrologist to determine location, lithology, relief, climate, and tectonic activity of the source area; to deduce the character of the environment of deposition; to determine the cause for changes in thickness or lithology; and to correlate beds precisely by mineral work. Sedimentary studies are also vital in prospecting for economic mineral reserves, especially as new deposits become harder to locate. Study of sediments is being pursued intensely by oil companies, phosphate, uranium, and iron mining companies in order to locate new deposits and explain the origin of those already known.

Fundamental Classification of Sedimentary Rocks. Sediments consist fundamentally of three components, which may be mixed in nearly all proportions: (1) Terrigenous components, (2) Allochemical components, and (3) Orthochemical components.

a. Terrigenous components are those substances derived from erosion of a land area outside the basin of deposition, and carried into the basin as solids; examples: quartz or feldspar sand, heavy minerals, clay minerals, chert or limestone pebbles derived from erosion of older rock outcrops.

b. Allochemical constituents (Greek: "allo" meaning different from normal) are those substances precipitated from solution within the basin of deposition but which are "abnormal" chemical precipitates because in general they have been later moved as solids within the basin; they have a higher degree of organization than simple precipitates. Examples: broken or whole shells, oolites, calcareous fecal pellets, or fragments of penecontemporaneous carbonate sediment torn up and reworked to form pebbles.

c. Orthochemical constituents (Greek: "ortho" meaning proper or true) are "normal" chemical precipitates in the customary sense of the word. They are produced chemically within the basin and show little or no evidence of significant transportation or aggregation into more complex entities. Examples: microcrystalline calcite or dolomite ooze, probably some evaporites, calcite or quartz porefillings in sandstones, replacement minerals.

Classes (b) and (c) are collectively referred to as "Chemical" constituents; classes (a) and (b) can be collectively termed "Fragmental." Some people use "detrital" or "clastic" as equivalent to "terrigenous"; other people use "detrital" or "clastic" as a collective term including both "terrigenous" and "allochemical" above.

Sedimentary rocks are divided into five basic classes based on the proportions of these three fundamental end members, as shown in the triangular diagram:

TERRIGENOUS

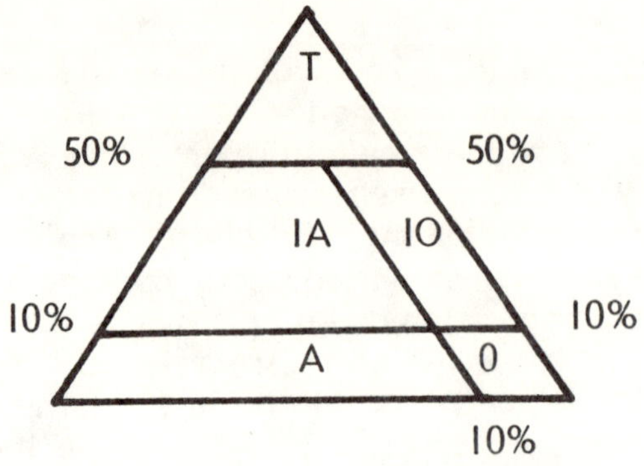

ALLOCHEMICAL ORTHOCHEMICAL

CHEMICAL

T. Terrigenous Rocks. Example: most mudrocks, sandstones, and conglomerates. Comprise 65% to 75% of the stratigraphic section. Most terrigenous rocks occur in the shaded area.

IA. Impure Allochemical Rocks. Example: very fossiliferous shales; sandy fossiliferous or oolitic limestones. Comprise 10-15% of the stratigraphic section.

IO. Impure Orthochemical Rocks. Example: clayey microcrystalline limestones. Comprise 2-5% of the stratigraphic section.

A. Allochemical Rocks. Example: fossiliferous, oolitic, pellet or intraclastic limestones or dolomites. Comprise 8-15% of the stratigraphic section.

O. Orthochemical Rocks. Example: microcrystalline limestone or dolomite; anhydrite; chert. Comprise 2-8% of the stratigraphic section.

Collectively, "IA" and "IO" are classed as Impure Chemical Rocks, and "A" and "O" as Pure Chemical Rocks.

PROPERTIES OF SEDIMENTARY ROCKS

Grain Size

Quantitative measurement of such grain size parameters as size and sorting is required for precise work. To measure grain size, one must first choose a scale. Inasmuch as nature apparently favors ratio scales over arithmetic scales, the grain size scale, devised by Udden, is based on a constant ratio of 2 between successive classes; names for the class intervals were modified by Wentworth. Krumbein devised the phi (ϕ) scale as a logarithmic transformation of the Wentworth scale, and modern data is nearly always stated in ϕ terms because mathematical computations are much simplified.

Grain size of particles larger than several centimeters is usually determined by direct measurement with calipers or meter sticks; particles down to about 4ϕ (0.062 mm) are analyzed by screening; and silts and clays (fine than 4ϕ) are analyzed by pipette or hydrometer, utilizing differential settling rates in water. Sands can also be measured by petrographic microscope or by various settling devices.

Results of grain-size analysis may be plotted as histograms, cumulative curves or frequency curves. Curve data is summarized by means of mathematical parameters allowing ready comparison between samples. As measures of average size the median, mode, and mean are frequently used; as a measure of sorting the standard deviation is best. Skewness and kurtosis are also useful parameters for some work.

Geological significance of the measures is not fully known. Many generalizations have been made on far too little evidence, but significant studies are now going on. As of our present state of knowledge, we can make a few educated guesses about the meaning of grain-size statistics.

The significance of mean grain size is not yet well enough known to make any positive statements; volumes of data on recent sediments must be collected before we can say anything really meaingful. To be sure there is a certain correlation of grain size with environments (see page 107)--e.g. you usually do not find conglomerates in swamps or silts on beaches--but there is a great deal of overlapping. These questions can only be solved by integration of size analysis results with shape study, detailed field mapping, study of sedimentary structures, fossils, thickness pattern changes, etc. It is extremely risky to postulate anything based on size analysis from one sample or one thin section, unless that sample is known to be representative of a large thickness of section. One thing that is an especially common error is the idea that if a sediment is fine it must be far from the source, while if it is coarse it must be near to a source. Many studies of individual environments show sediments getting finer away from the source but these changes are so varied that they can be deciphered only by extensive field and laboratory work. For example, strong rivers may carry large pebbles a thousand or so miles away from their source, while on the other hand the finest clays and silts are found in playa lakes a matter of a few miles from encircling rugged mountains. Grain size depends largely on the current strength of the local environment (together with size of available particles), not on distance. Flood plain clays may lie immediately adjacent to course fluvial gravels, both being the very same distance from their source. One must work up a large number of samples before anything much can be said about the significance of grain size. Still, it is an important descriptive property, and only by collecting data on grain size will we be able to learn the meaning of it.

Mean size is a function of (1) the size range of available materials and (2) amount of energy imparted to the sediment which depends on current velocity or turbulence of the transporting medium. If a coastline is made up of out-crops of soft, fine-grained sands, then no matter how powerful the waves are no sediments coarser than the fine sands will ever be found on the beach. If a coastline is made up of well-jointed, hard rocks which occasionally tumble down during rains, then the beach sediment will be coarse no matter how gentle the waves of the water body. Once the limitations of source material are understood, though, one can apply the rule that sediments generally become finer in the direction of transport; this is the case with most river sands, beach sands, spits and bars. This is largely the result not of abrasion, but of selective sorting whereby the smaller grains outrun the larger and heavier ones in a downcurrent direction. Pettijohn and students have made excellent use of maximum pebble size in predicting distance of transport quantitatively. Sediments usually become finer with decrease in energy of the transporting medium; thus, where wave action is dominant sediments become finer in deeper water because in deep water the action of waves on the sea bottom is slight, whereas this turbulence is at a maximum in shallow waters at the breaker zone. Where current action dominates, particularly in tidal channels, coarses sediments occur in deeper waters, largely because of scour. Research is needed to quantify these changes so that the rate of grain-size change with depth can be correlated with wave energy expenditure or other environmental factors.

Sorting is another measure which is poorly understood. It depends on at least three major factors: (1) Size range of the material supplied to the environment--obviously, if waves are attacking a coastline composed of glacial till with everything from clay to room-sized boulders, the beach sediments here will not be very well sorted; or if a turbulent river is running through outcrops of a friable well-sorted Tertiary sand then the river bars will be well sorted. (2) Type of deposition--"bean spreading", with currents working over thin sheets of grains continuously (as in the swash and backwash of a beach) will give better sorting than the "city-dump" deposition in which sediments are dumped down the front of an advancing series of crossbeds and then rapidly buried by more sediment. (3) Current characteristics--currents of relatively constant strength whether low or high, will give better sorting than currents which fluctuate rapidly from almost slack to violent. Also very weak currents do not sort grains well, neither do very strong currents. There is an optimum current velocity or degree of turbulence which produced best sorting. For best sorting, then, currents must be of intermediate strength and also be of constant strength. (4) Time--rate of supply of detritus compared with efficiency of the sorting agent. Beach sediments where waves are attacking continually caving cliffs, or are battling great loads of detritus brought to the shore by vigorous rivers, will be generally more poorly sorted than beaches on a flat, stable coast receiving little sediment influx.

It is probable that in every environment, sorting is strongly dependent on grain size. This can be evaluated by making a scatter plot of mean size versus sorting (standard deviation). In making many of these plots, a master trend seems to stand revealed: the best sorted sediments are usually those with mean sizes of about 2 to 3ϕ (fine sand) (Griffiths; Inman). As one measures coarser sediments, sorting worsens until those sediments with a mean size of 0 to -1ϕ (1 to 2 mm) show the poorest sorting values. From here sorting improves again into the gravel ranges (-3 to -5ϕ), and some gravels are as well sorted as the best-sorted sands (Folk and Ward). Followed from fine sand into finer sediments, the sorting worsens so that sediments with a mean size of 6 to 8ϕ (fine silts) have the poorest sorting values, then sorting gradually improves into the pure clay range (10ϕ). Thus the general size vs. sorting trend is a distorted sine curve of two cycles. Work so far indicates that the apparent reason for this is that Nature produces three basic populations of detrital grains to rivers and beaches (Wentworth). (1) A pebble population, resulting from massive rocks that undergo blocky breakage along joint or bedding planes, e.g. fresh granite or metaquartzite outcrops, limestone or chert beds, vein quartz masses. The initial size of the pebbles probably depends on spacing of the joint or bedding planes. (2) A sand-coarse silt population, representing the stable residual products liberated from weathering of granular rocks like granite, schist, phyllite, metaquartzite or older sandstones whose grains were derived ultimately from one of these sources. The initial size of the sand or silt grains corresponds roughly to the original size of the quartz or feldspar crystal units in the disintegrating parent rocks. (3) A clay population, representing the reaction products of chemical decay of unstable minerals in soil, hence very fine grained. Clays may also be derived from erosion of older shales or slates whose grain size was fixed by the same soil-forming process in their ultimate source areas. Under this hypothesis, a granite undergoing erosion in a humid climate and with moderate relief should produce (1) pebbles of granite or vein quartz from vigorous erosion and plucking of joint blocks along the stream banks, (2) sand-size quartz grains, and (3) clay particles, both as products formed in the soils during weathering.

Because of the relative scarcity in nature of granule-coarse sand (0 to -2ϕ) particles, and fine silt [6 to 8ϕ grains, sediments with <u>mean sizes</u> in these ranges must be a mixture of either (1) sand with pebbles, or (2) sand or coarse silt with clay, hence will be more poorly sorted than the pure end-members (pure gravel, sand, or clay)]. This is believed to be the explanation of the sinusoidal sorting vs. size trend. Of course exceptions to this exist, e. g. in disintegration of a coarse-grained granite or a very fine phyllite which might liberate abundant quartz grains in these normally-rare sizes. If a source area liberates grains abundantly over a wide range of sizes, sorting will remain nearly constant over that size range (Blatt) and no sinusoidal relation will be produced. Shea (1974 JSP) denies existence of "gaps" in natural particle sizes.

Although it appears that all sediments (except glacial tills) follow this sinusoidal relation, there is some differentiation between environments. It is believed that given the same source material, a beach will produce better sorting values for each size than will a river; both will produce sinusoidal trends, but the beach samples will have better sorting values all along the trend because of the "bean spreading" type of deposition. Considering only the range of sands with mean sizes between 1ϕ and 3ϕ., most beach sands so far measured here have sorting (σ_I) values between .25-.50ϕ, while most river sands have values of .35-1.00ϕ. Thus there is some overlap, and of course there are some notable exceptions; beach sands formed off caving cliffs are more poorly sorted because the continual supply of poorly sorted detritus is more than the waves can take care of, and rivers whose source is a well-sorted ancient beach or marine sand will have well-sorted sediments. Coastal dune sands tend to be slightly better sorted than associated beaches, though the difference is very slight, but inland desert dunes are more poorly sorted than beaches. Near shore marine sands are sometimes more poorly sorted than corresponding beaches, but sometimes are better sorted if longshore currents are effective. Flood-plain, alluvial fan, and offshore marine sediments are still more poorly sorted although this subject is very poorly known and needs a great deal more data. Beach gravels between 0 ϕ and - 8 ϕ, whether made of granite, coral, etc., have characteristic sorting values of 0.4 -0.6 ϕ (if they are composed mainly of one type of pebble); there seems to be no difference in sorting between beaches with gentle wave action vs. those with vigorous surf. (Folk).

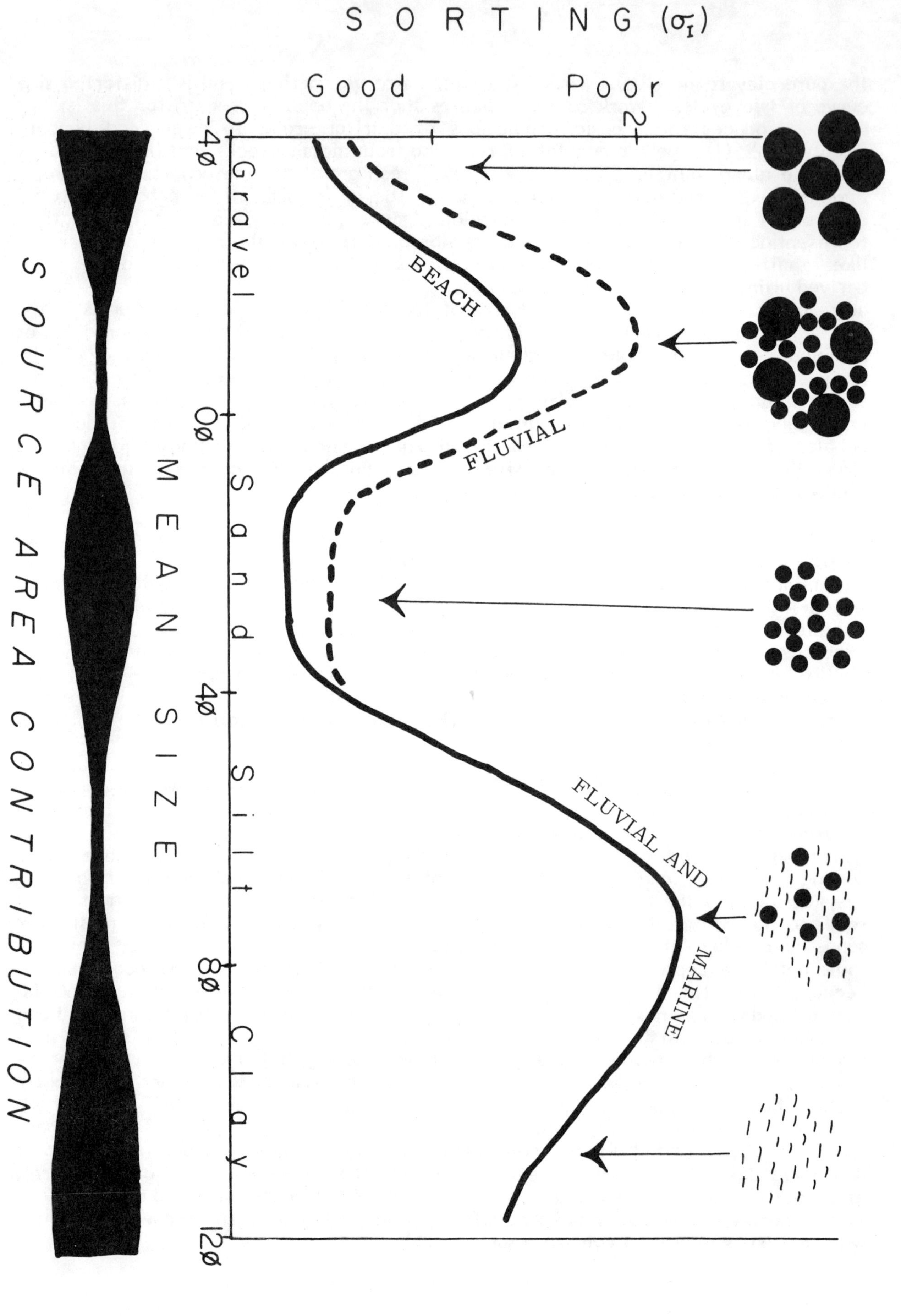

SORTING (σ_I)
Good
Poor
0
1
2
-4ø
0ø
4ø
8ø
12ø
Gravel
Sand
Silt
Clay
BEACH
FLUVIAL
FLUVIAL AND
MARINE
MEAN SIZE
SOURCE AREA CONTRIBUTION

Skewness and kurtosis tell how closely the grain-size distribution approaches the normal Gaussian probability curve, and the more extreme the values the more non-normal the size curve. It has been found that single-source sediments (e.g. most beach sands, aeolian sands, etc.) tend to have fairly normal curves, while sediments from multiple sources (such as mixtures of beach sands with lagoonal clays, or river sands with locally-derived pebbles) show pronouned skewness and kurtosis. Bimodal sediments exhibit extreme skewness and kurtosis values; although the pure end-members of such mixtures have nearly normal curves, sediments consisting dominantly of one end member with only a small amount of the other end member are extremely leptokurtic and skewed, the sign of the skewness depending on which end member dominates; sediments consisting of subequal amounts of the two end-members are extremely platykurtic. (Folk and Ward). Plots of skewness against kurtosis are a promising clue to environmental differentiation, for example on Mustang Island (Mason) beaches give nearly normal curves, dunes are positively-skewed mesokurtic, and aeolian flats are positively-skewed leptokurtic. Friedman showed that dunes tend to be positive skewed and beaches negative skewed for many areas all over the Earth, but Hayes showed on Padre Island that this is often modified by source of supply. Eolian deflation sediments are commonly bimodal.

Fluvial environments consisting chiefly of traction load (coarse) with some infiltrated suspension load (finer grains) are commonly positive-skewed leptokurtic; glacial marine clays with ice-ratified pebbles are negative-skewed, etc. It would be emphasized that faulty sampling may also cause erroneous skewness and kurtosis values, if a worker samples two adjoining layers of different size - i. e., a gravel streak in the sand. Each layer should be sampled separately.

Size analysis has been used practically in correlation of formations; in determining if a sand will contain oil, gas or water (Griffiths); in determining direction of sediment transport; and an intensive study is being made to determine if characteristic grain size distributions are associated with certain modern environments of sedimentation, so that such deposits may be identified by analysis of ancient sediments in the stratigraphic column. Furthermore many physical properties of sediments such as porosity, permeability, or firmness (Krumbein) are dependent on the grain size.

Particle Morphology

Under the broad term "particle morphology" are included at least four concepts. Listed approximately in decreasing order of magnitude, these are (1) form, (2) sphericity, (3) roundness, and (4) surface features.

Form is a measure of the relation between the three dimensions of an object, and thus particles may be classed quantitatively as compact (or equidimensional), elongated (or rodlike) and platy (or disclike), with several intermediate categories, by plotting the dimensions on a triangular graph (Sneed and Folk: see p. 9).

Sphericity is a property whose definition is simple, but which can be measured in numerous very different ways. It states quantitatively how nearly equal the three dimensions of an object are. C. K. Wentworth made the first quantitative study of shapes. Later, Waddell defined sphericity as

$$\sqrt[3]{\frac{V_p}{V_{cs}}}$$

where V_p is the actual volume of the particle (measured by immersion in water) and V_{cs} is the volume of the circumscribing sphere (the smallest sphere that will just enclose the particle); actually the diameter of this sphere is considered as equal to the longest dimension of the particle. A sphere has a sphericity of 1.00; most pebbles or sand grains have a sphericity of about 0.6-0.7 when measured by this system. One can get an approximation of this measure by the formula,

$$\sqrt[3]{\frac{LIS}{L^3}}$$

where L, I and S represent the long, intermediate and short dimensions respectively (Krumbein).

The above formula, although in common use today, does not indicate how the particle behaves upon settling in a fluid medium and is thus rather unsatisfactory. Actually, a rod settles faster than a disk of the same volume, but Waddell's formula would have us believe the opposite. A sphericity value which shows better the behavior of a particle during transport is the Maximum Projection Sphericity (Sneed and Folk, J. Geol. 1958), given by the formula

$$\sqrt[3]{\frac{S^2}{LI}} .$$

Particles tend to settle with the maximum projection area (the plane of the L and I axes) perpendicular to the direction of motion and hence resisting the movement of the particle. This formula compares the maximum projection area of a given particle with the maximum projection area of a sphere of the same volume; thus if a pebble has a sphericity of 0.6 it means that a sphere of the same volume would have a maximum projection area only 0.6 as large as that of the pebble. Consequently the pebble would settle about 0.6 as fast as the sphere because of the increased surface area resisting downward motion. The derivation follows: assuming the particle to be a triaxial ellipsoid, the maximum projection area of the particle is $\pi/4$ (LI). The volume of the particle is $\pi/6$ (LIS). Hence the volume of the equivalent sphere will also be $\pi/6$ (LIS). The general formula for the volume of a sphere is $\pi/6\ d^3$. Therefore, in this example, $d^3 = LIS$ and the diameter of the equivalent sphere, d, will equal

$$\sqrt[3]{LIS} .$$

The maximum projection area of this sphere will equal

$$\frac{\pi}{4}\left(\sqrt[3]{LIS}\right)^2 .$$

The maximum projection sphericity then equals

$$\frac{\frac{\pi}{4}\left(\sqrt[3]{IIS}\right)^2}{\frac{\pi}{4}(LI)} ,$$

which reduces to

$$\sqrt[3]{\frac{L^2 I^2 S^2}{L^3 I^3}} = \sqrt[3]{\frac{S^2}{LI}} = \Psi_p .$$

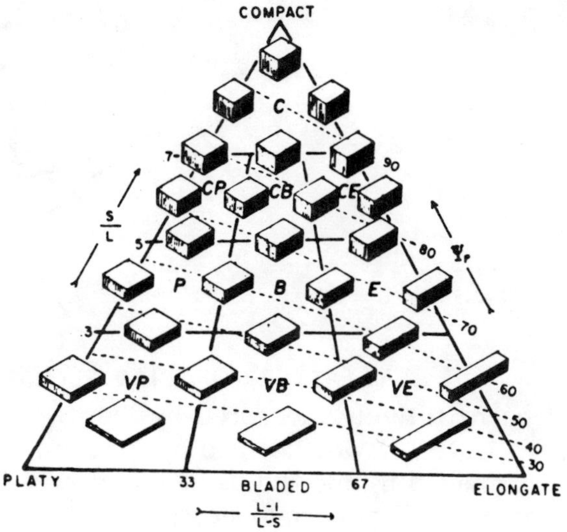

Form triangle. Shapes of particles falling at various points on the triangle are illustrated by a series of blocks with axes of the correct ratio; all blocks have the same volume. Independence of the concepts of sphericity and form may be demonstrated by following an isosphericity contour from the disklike extreme at the left to the rodlike extreme at the right.

For two-dimensional purposes (as in thin sections) two other "sphericity" measurements have been applied. Riley Sphericity is given as

$$\sqrt[2]{\frac{D_i}{D_c}}$$

where D_c is the diameter of the smallest circumscribing circle and D_i is the diameter of the largest inscribed circle. These can be easily measured by a celluloid scale ruled off in a series of closely-spaced concentric circles of known diameter, which can then be placed over the sand grain image. Another measure is Elongation which is simply width (actually least projection width) over length measured by rectangular grid. This is probably the most satisfactory two dimensional measure (Griffiths; Dapples and Rominger).

Although individual grains may have widely varying W/L values, sample means (obtained by counting 100 quartz grains in one thin section, for example) show a much more restricted range. Measurement of many sandstones has suggested the following scale: under .60, very elongate; .60-.63, elongate; .63-.66, subelongate; .66-.69, intermediate shape; .69-72, subequant; .72-.75, equant; and over .75, very equant.

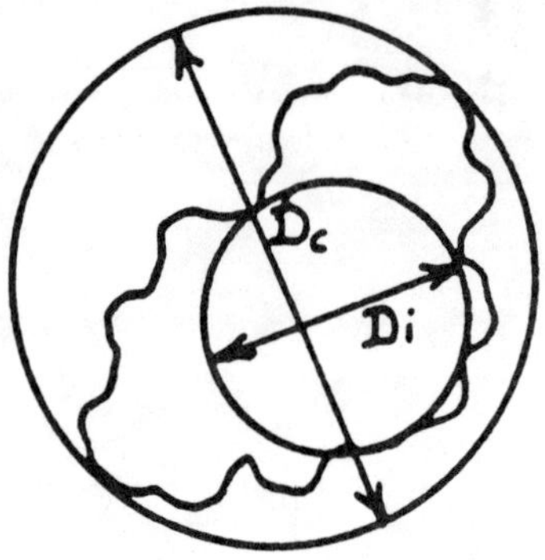

RILEY SPHERICITY

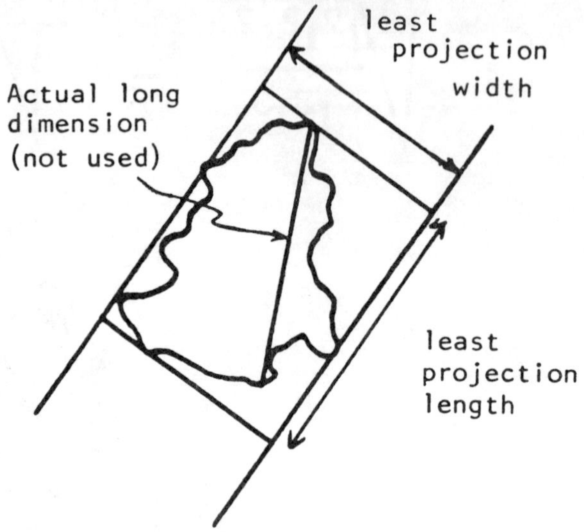

LEAST PROJECTION ELONGATION

Roundness was first quantitatively measured by Wentworth, who used the curvature of the sharpest corner. Later, it was defined by Waddell as the average radius of curvature of all the corners divided by the radius of the largest inscribed circle. This is impractical to measure, though, and now roundness values are obtained by comparison with photographic charts for sand grains (Powers). A perfect ball has a roundness of 1.0; most sand grains have roundnesses about 0.3-0.4 on the Waddell scale. Use of the Powers roundness images for sand grains is facilitated by a logarithmic (rho, ρ) scale in which the limits of the very angular class are taken as 0.0-1.0, of angular 1.0-2.0, subangular 2.0-3.0, subround 3.0-4.0, round 4.0-5.0, and very round 5.0-6.0ρ (Folk). On this scale, perfect balls have a roundness of 6.0ρ and most sand grains have average roundness about 2.5ρ (subangular).

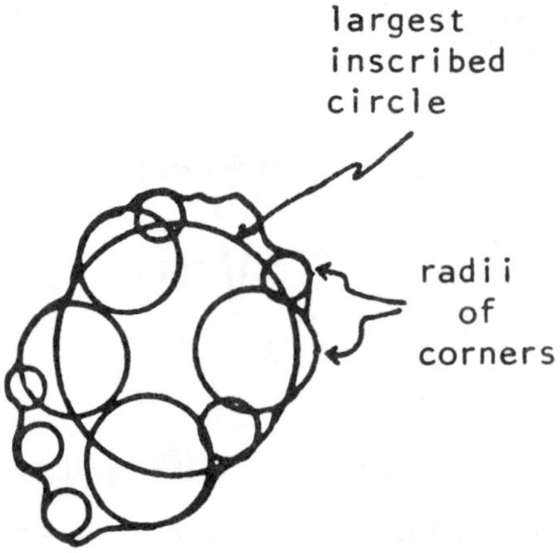

The concept of roundness sorting (uniformity of roundness) may be expressed by the roundness standard deviation, σρ. This can be determined by plotting the roundness data as a cumulative curve and determine σρ by the intercept method, as is done in grain size curves. Plotting of these values for many samples gives the following limits for roundness standard deviation: under 0.60, very good roundness sorting; 0.60-0.80, good; 0.80-1.00, moderate; 1.00-1.20, poor; over 1.20, very poor roundness sorting. St. Peter sand, uniformly well rounded, has σρ under 0.60; the average value for Recent sands is about 0.90; and the Colorado River at Austin, with its mixture of well-rounded, Cambrian and Cretaceous quartz with freshly broken Llano granitic quartz has a $\sigma\rho$ of 1.30.

Pebble roundness can be measured more quantitatively by dividing the radius of curvature of the sharpest single corner, by the radius of the largest inscribed circle (see Dobkins and folk, 1970 J. S. P.)

<u>Surface Features.</u> As yet no way has been developed to measure these quantitatively. Frosted surfaces are noticeable chiefly on round grains, although not all round grains are frosted. Frosting may be caused by chemical etching (as by dolomite replacement), by incipient tiny quartz overgrowths, or by aeolian abrasion (where the frosting is due to minute crescentic percussion marks caused by the greater impact and velocity in air). These may be distinguished by examining the grain with petrographic microscope in water. Polished surfaces are caused by a fine smoothing of the tiny irregularities and are ascribed to rubbing of the grains in water; thus they are the first stage in rounding of aqueous sands. They are supposedly common on beaches but much less so in neritic or river sands where the grains are not rolled back and forth so much; but there is considerable question as to the validity of this criterion, and some polish may form chemically (Kuenen). Much volcanic phenocryst quartz has naturally polished faces. Unpolished surfaces have neither frosting or polish, and are dull or jagged because of tiny angular irregularities or fresh fracture surfaces. They are chiefly found in river or neritic sands and indicate lack of much abrasive action. Percussion marks are found on pebbles (especially chert or quartzite) and indicate high-velocity impact. Glacial action sometimes produces scratches or striae on soft pebbles like limestone. The entire subject of surface features needs much more research to find out what happens in various environments and to what extent the observed surface features are a function of grain size.

As a possible explanation for the conflicting data on surface features and environment, the following completely untested wild idea is put forth: all abrasional environments produce all types of surface features, but on different sizes of particles. In each environment the coarsest grains are frosted (probably by percussion marks), intermediate grains are polished, and the finest grains are dull (unmodified). The size ranges over which these surface features develop on quartz varies with the environment, perhaps as below:

size	-4 -2 0 2 4
Neritic	D U L L
River	F r o s t e d P o l i s h e d D u l l
Beach	F r o s t e d P o l i s h e d D u l l
Dune	F r o s t e d Polished Dull

The most common size inspected for surface features is about medium sand, which would explain the feeling that most dunes are frosted, most beach sands polished, and river sands dull. Of course, rate of deposition, time available, and rate of supply add complications, and some surface features are formed chemically. Mature desert quartz often has a greasy luster caused by a minutely pimply deposition of quartz "turtle-skin" overgrowths.

As a rule, it is difficult to study surface features in ancient sediments because of cementation. An attempt may be made by cleaning the grains thoroughly in warm HCl to remove carbonates and iron stains. They should be studied under highest power of the binocular microscope, on a black surface and under strong light. After this, the grain should be mounted in water and examined with petrographic microscope. Electron microscopy has made great contributions to study of grain surface features (Krinsley), and intimate details of grain history can be studied.

Graphic and Statistical Analysis of Shape Data. Graphic or statistical analysis is necessary to show (1) the variation of grain morphology with size, which is nearly always present, and (2) environmental, spatial, or stratigraphic differences between samples. In comparing a set of samples, one should choose the same size interval throughout for his analyses, because coarser grains are usually better rounded, and show different surface features. However on some of the samples, shape study should be made on all sizes. This is called an H pattern of sampling.

If a significant difference in morphology is found, (for example if beach sands in an area are more angular than dune sands), you must always consider whether it is due to (1) an actual difference in the processes going on (that the dunes are rounding more effectively) or whether (2) it is simply the result of selective sorting, where e.g. rounded and more spherical grains are left behind by wind currents which selectively sweep away the more angular and less spherical grains.

To determine if a difference in Form is present between two sets of samples, one can (1) using a moving circular mask, contour the points on the triangular diagram as is done in contouring joint or petrofabric diagrams; (2) use the X^2 test by counting the number of particles in each shape "cell" and comparing the results; (3) obtain by counting a median shape for each set of data; (4) superimpose a sheet of transparent triangular-ruled graph paper over the data, and assign each point a "percent elongation" and a "percent platiness" by measuring its distance from the base line of the triangle; this data may be treated by finding its mean and standard deviation and comparing by the t test.

Sphericity differences between samples can be evaluated by finding a mean and standard deviation for each set of particles and comparing by means of the t test.

Roundness is analyzed by counting a large number of particles using comparison charts; then, use the log transformations of the roundness values (ρ scale) and compute means (average roundness) and standard deviations (roundness sorting), then compare by the t test. The roundness data may also be plotted by means of cumulative curves if probability paper is used. In doing a set of samples from two or more environments, formations, or localities, it is well to have an associate mix the samples up so that you do not know which set you are working on when you count a sample; thus you avoid a bias and the data is more "honest."

Surface features may be compared by counting the number of frosted, polished grains etc. and comparing by the X^2 test. Again, an associate should mix the samples up so you do not know which sets you are counting, in order to avoid bias.

Significance of Grain Morphology

Form and Sphericity are the result of (1) Structure (internal properties, inherited from the source; (2) Process (work of the depositional environment e.g. glacier, river or beach; and (3) Stage (length of time available for modifying the particle). This is the concept drawn from W. M. Davis, who used it for landscape interpretation. Concerning Structure, bedding, schistosity and cleavage tend to make particles wear to discoids; directional hardness may have some effect in minerals like quartz and kyanite. Original shape of the particle may be somewhat retained, as in shapes of joint blocks, or in platy quartz derived from some schists.

The effects of Process and Stage are complex, and there is insufficient data on this aspect. In studying pebble shapes, most workers have used a grand mixture of rocks, including schists, sandstones, thin-bedded limestones, etc. Of course such data is useless; to get valid environmental data one must use isotropic wearing rocks such as granite, basalt, or quartz. Further, the same geologic process (e.g. surf action) may work to different ends with different sized pebbles, so pebble size must be carefully controlled (i.e. use the same size pebbles for the entire study).

A study of pebbles in the Colorado River was made by Sneed. He found that the main control on sphericity is pebble lithology, with chert and quartz having much higher sphericity than limestone. Smaller pebbles develop higher sphericity than larger ones with long transport. Going downstream, sphericity of quartz increased slightly, limestone stayed constant, and chert decreased. He found a weak tendency for the Colorado River to produce rod-like pebbles in the larger sizes.

Dobkins and Folk (1970 J. S. P.) found on Tahiti beaches, using uniformly-wearing basalt, that beach pebble roundness averaged .52 while river pebbles were .38. Beach pebble were oblate with very low sphericity (.60), while fluvial pebbles averaged much higher sphericity, .68. Relations are complicated by wave height, substrate character (sandy vs. pebbly), and pebble size; nevertheless, a sphericity line of .65 appears to be an excellent splitter between beach and fluvial pebble suites, providing isotropic rocks are used. Apparently this is caused by the sliding action of the surf.

Roundness probably results from the chipping or rubbing of very minute particles from projecting areas of the sand grains or pebbles. Solution is not held to be an important factor in producing roundness, though some dispute this (Crook, 1968). Impact fracturing of grains (i.e. breaking them into several subequally-sized chunks) is not an important process except possibly in mountain torrents; "normal" rivers carry few fractured pebbles, and beach or river sands only rarely contain rounded and refractured grains. Rounded pebbles may break along hidden joints if exposed to long weathering. The "roundability" of a particular mineral or rock fragment depends upon its hardness (softer grains rounding faster), and the cleavage or toughness (large grains with good cleavage tend to fracture rather than round; brittle pebbles like chert also tend to fracture readily whereas quartz pebbles will round). On the Colordado River, limestone pebbles attain their maximum roundness in only a few miles of transport and thereafter do not get any more rounded. Quartz also becomes well rounded but at a much slower rate, equalling the roundness of limestone after about 150 miles; chert shows only a slight increase in roundness in over 200 miles of transport. Coarser grains round easier than finer ones, because they hit with greater impact and also tend to roll along the surface while finer ones may be carried in suspension. Aeolian sands round faster than aqueous sands because the grains have a greater differential density in air, therefore hit harder; also they are not cushioned by a water film. In experiments by Kuenen, wind-blown quartz rounded much more rapidly than water-transported grains. Beach sands also round faster than river sands because on a beach the grains are rolled back and forth repeatedly. It is thought that little or no rounding of sand grains takes place in rivers; pebbles get rounded rapidly in rivers, however. To get rounded grains takes a tremendous time and a very large ependiture of energy. For example, the beach sands of the Gulf Coast show very little rounding taking place simply because the shoreline fluctuates so rapidly that time is not adequate. Balasz and Klein report rapid rounding in a "merry-go-round" tidal bar.

In studying roundness watch for two things: (1) an abnormal relation between roundness and mineral hardness, e.g. if tourmaline is round and softer hornblende is angular it means there is a multiple source area; and (2) an abnormal relation between size and rounding, e.g. if coarse grains are angular and fine ones are round it again usually means a multiple source with the angular grains being primary, the rounder ones coming from reworked sediments. Also, poor roundness sorting (i.e. within the same grade size there are both very rounded and very angular grains) indicates a multiple source. To determine how much rounding is taking place in the last site of deposition, look for the most angular grains; for example a sand consisting of 70% well rounded grains and 30% angular grains indicates little or no rounded grains have simply been inherited from older sediments. Perhaps the 16 percentile of roundness is the best parameter to evaluate the present rate of rounding.

Effect of transportation on grain size and morphology. Transportation does reduce the size of pebbles through chipping or rubbing and occasionally through fracturing; but it is thought that very little size reduction of sand-sized quartz is effected through transport. The differences in size between deposits are chiefly due to selective sorting (where the finer particles, which travel in almost all currents, outrun the coarser particles, which can travel only in the strong currents), rather than to abrasion. Thus one cannot say that a very fine sand has been abraded longer than a fine sand: simply it has been carried farther or deposited by weaker curents. The effect of abrasion on sphericity of sand is slight but noticeable. Crushed quartz and many angular sands have W/L values of about .60-.64; very well rounded sands have W/L of over .70. Selective sorting will also produce form and sphericity differences between samples.

It can be certainly stated that abrasion does cause rounding of sand grains (albeit very slowly), and that it even more rapidly will produce polish. Thus the smallest order features are affected first: considering sand grains in water, starting initially with crushed quartz, the first thing that happens is that the grains become polished; after much more abrasion, they become rounded; after an extreme length of time they begin to attain higher sphericity values; and still later their size is reduced. Really these processes are all taking place at once, and the above list simply gives the rates at which these changes are being effected. Surface features and, secondarily, roundness are the important clues to the latest environment in which the sand was deposited; sphericity and form are the clues to the earliest environment in which the sand was formed, namely source rock.

COLLECTION AND PREPARATION OF SAMPLES FOR ANALYSIS

Sampling

There are two methods of sampling, the channel sample and the spot sample. The channel sample is useful when you are trying to get average properties over a large stratigraphic interval. For example, if you are making a channel sample of a sand exposed in a 20-foot cliff, you can dig a trench the full height of the cliff and take a continuous sample all the way up, mix it up thoroughly and analyze the whole thing together, or you can take small samples every foot, mix them up thoroughly and analyze the mixture. This method is good for determining economic value, for example if you would like to find out how much iron or how much clay mineral is present in the whole cliff. But it is absolutely useless for determining origin or sedimentational conditions, since you are analyzing a mixture of many layers deposited under varying conditions. We will therefore use spot sampling. To take a spot sample, one selects a representative small area and takes a sample of just one bed that appears homogeneous--i.e., you sample just one "sedimentation unit." For example, if an outcrop consists of two-inch beds of coarse sand alternating with one-inch beds of fine sand, your spot sample will include only one of these beds, trying to get it pure without any admixture of the alternate bed. The smaller your sample is, the more nearly will it represent sedimentational conditions; for sands, 50-100 grams is probably adequate (this is a volume of 2 to 4 cubic inches); for gravels, a larger sample is required to get a representative amount of pebbles, thus usually 1000 grams or more is needed (50 to 100 cubic inches); for silts and clays, you may sample as small as 20 to 50 grams. Just remember to sample only one layer of sediment if possible, excluding any coarser or finer layers that lie adjacent.

If you have the problem of sampling a sedimentary body, be it a recent sand bar or a Cambrian shale, first decide on the extent of the unit to be sampled (which may be either area, linear distance along an outcrop, or stratigraphic thickness) and then on the number of samples you want to take. Dividing these two numbers gives you the spacing of the samples. If you are going to take only a few samples, it is a general rule to take representative ones (i.e., sample the most common or typical beds exposed). If you are going to take a large number of samples, the ideal way to approach it is as follows: (1) after dividing the extent of the unit by the number of samples, set up an equispaced grid of sampling stations through the unit. For example if the cliff is 20 feet high and you want to take about 7 samples, then 20/7 = 3 and you space your main sample locations at 1.5, 4.5, 7.5, 10.5, 13.5, 16.5, 19.5 feet and sample the material at those precise points. By doing this according to a rigid grid spacing, you are pretty sure to get representative samples of the typical rock. In addition to the samples spaced on the grid, you should take extra samples of different or peculiar layers that may have been missed on the grid; in the example above, if a conglomerate layer had occurred at 12 feet you would take an extra sample of that even if you did not land on the grid, and you'd end up with 8 samples of the outcrop. These extra samples should be especially labeled inasmuch as they are somewhat unique and not representative of the outcrop. The whole purpose of the grid sampling method is to avoid the error of non-representative samples; if one were to sample a batholith, for example, one would not collect 40 samples of pegmatites and only 5 samples of legitimate granite no matter how monotonous and similar the granite appeared; one would collect chiefly granite samples with only a few pegmatites. Above all, use your head and in the samples you collect try to end up with the number of samples of each rock type in proportion with the amount of that rock type in the exposure.

In detailed work on sedimentary units it is necessary to obtain samples showing the complete range of grain size present in the unit--i.e. you should search for and sample the coarsest and finest beds obtainable as well as getting samples typical of the unit, for only in this way can the interrelation between size, sorting, shape, and mineral composition be studied.

Preparation of Samples for Grain-Size Analysis

The purpose of grain size analysis is to obtain the grain size of the clastic particles as they were deposited. This is often difficult because (1) clastic sand grains may acquire overgrowths, or be cemented into tough aggregates to form hard sandstones; (2) chemically-precipitated materials may be introduced into the rock which, if not removed, give erroneous size values because we are trying to measure size of the clastic particles, not size of the cementing material; and (3) clay minerals because of their flaky character and surface electrical charges tend to cluster in lumps. The idea of disaggregation and dispersion is to separate all of the individual grains without smashing any of them, and to remove all chemically-precipitated substances. To avoid smashing grains as much as possible, one should start with gentle treatments and gradually work up to the more severe treatments only if the gentle ones fail.

a. Disaggregation of Sands and Sandstones.

1. Unconsolidated Sediments. The sample should be dried, then placed on a large sheet of glazed paper and crushed with the fingers. Spread the sand out and examine with hand lens or binocular microscope to see that all aggregates are crushed. Then gently rub these aggregates with the fingers or run over the sand with a small rolling-pin. Alternately, the sand may be poured into a mortar and gently pounded with a rubber cork. If there are a good many aggregates, it is often helpful to select a screen just larger than the size of most of the individual grains of the sediment, then run the whole sample through it; nearly all the grains passing through will then represent single grains and most of the particles remaining on the screen will be aggregates. In this way you can concentrate your efforts on crushing the aggregates without wasting your energy on the single grains.

2. Weakly Consolidated Sediments. Place the sample in a mortar, and place a large sheet of paper under the mortar. First, try crushing with a rubber cork; if this doesn't work, try gentle pounding with an iron or porcelain mortar. Always use an up-and-down motion; never use a grinding motion as you will break the individual grains. Examine with binocular to check that aggregates are completely destroyed. Be careful not to splatter any sand out of the mortar.

3. Carbonate-cemented Rocks. With a mortar and pestle crush the sample to pea-size chunks or smaller. Place it in dilute hydrochloric acid until effervescence ceases (be sure the acid is still potent when effervescence ceases). If the sample is dolomite, it may be heated gently to hasten solution. Pour off the acid and wash the sample. If any fine clays are suspended in the acid after this treatment, pour the

acid and the wash water through a filter paper that you previously weighed, and place in an oven to dry. The weight of this dried material then should be added to the "pan" fraction of the sieve analysis. After the sand sample has been washed, dry it in the oven.

4. Ferruginous-cemented Rocks. Crush the sample to pea size and place in 50% HCl, warmed over a hot plate. Continue heating until the sand turns white. Pour off the acid (through a filter paper if "fines" are suspended in the liquid). Dry in an oven, and if it is necessary to do any more crushing, follow (1) or (2) above.

5. Siliceous-cemented Rocks. If the cement is opal or chalcedonic or microcrystalline quartz (i.e., chert), then warm concentrated KOH may work. If the cement is crystalline quartz there is no known chemical way to remove it and still leave the quartz grains. For rocks not too strongly cemented with quartz, try the pounding routine of (2) above. For quartzites, there is no satisfactory mechanical method of grain-size analysis, and it must be done in thin-section with a petrographic microscope.

6. Sands Bonded With a Little Clay. Place the sample in a wide dish with some water and rub with a cork until the clay is in suspension and grains are separated. The clay may then be removed by decantation or wet-sieving, and either weighed or analyzed by pipette or hydrometer. NEVER WASH THE CLAY DOWN THE DRAIN--ALWAYS WEIGH IT.

(Note: after you have done your sieve analysis, if you find an abnormally high percentage of aggregates (say over 25%) on any screen, that size fraction may be removed, recrushed, and rescreened. Otherwise the grain-size analysis is valueless).

If these methods fail, or for special types of cement, see Krumbein and Pettijohn (1938), Carver (1971), or Royse (1970).

b. Dispersion of Muds and Clays

Grain-size analysis of fine-grained materials is not very satisfactory and there are a great many unsolved problems. In the first place we are trying to measure the size of the individual particles, and to separate clay lumps into individual grains is very difficult. As you may see, the grain-size distribution we obtain on analysis may not really be a measure of the true size of the particles, but only tells how efficient our disaggregation has been. In general, one can rely on these analyses as giving a fairly true picture of the size-distribution down to diameters of 6 or 7 ϕ (.016 to .008 mm) but for sizes finer than this the analyses are often invalid. Below this size the analysis no longer measures true size of the particles, because the settling velocity is now affected greatly by the flaky shapes of the particles, degree of dispersion, electrical charges on the particles, etc. Two clay flakes of the same size but different compositions (e.g. kaolinite vs. montmorillonite) may settle at different rates because of these factors.

1. Dispersants. Clay flakes in distilled water are usually electrically charged. Most clays have a negatively-charged ionic lattice, which to attain electric neutrality must take up positively charged ions from the surrounding

solution (usually H+, but sometimes Na+, Ca++ or others). This leaves the surrounding solution swarming with unsatisfied negative ions (OH-, etc.) in the vicinity of the clay flakes. Thus when a clay flake with its surrounding "fog" of negative ions approaches another clay flake, also with a negative fog of ions around it, the two flakes repel each other. This is the state we try to maintain, because if the flakes repel each other then they will not aggregate into clumps, and we can then make a grain-size analysis on the individual grains. If however we add a strong electrolyte like HCl or NaCl to the solution, then the solution becomes a conductor and the negative swarm of ions leaves the clay crystal, and left unprotected the clay flakes attract each other and flocculate into chains and large clumps. Grain-size analysis is thereby made impossible. Therefore we must always use distilled water, never tap water, in these analyses, because tap water always contains small amounts of ions which cause flocculation. The best way to insure a good analysis is to add a small amount of certain dispersing chemicals (called peptisers) which prevent flocculation. These chemicals either build up the charges on the clay particles so that they repel each other more strongly, or else "plate' the particles with a protective and repulsive coating. The quantity of peptiser or dispersant used must be pretty exact because too little peptser can result in flocculation, and too much may also cause flocculation. To a large extent the grain size obtained by analysis depends on the type of dispersant, efficiency of stirring, etc. Some disperants are the following: (1) a few drops of conc. NH_4OH per liter of suspension; (2) .02N Na_2CO_3; (3) 0.01 N sodium oxalate; (4) 2.55gm/l sodium hexametaphosphate. The latter (Calgon) is best for most purposes. In all your operations using dispersants, keep the water you are using to mash or wash or suspend the particles in, always at the exact concentration of dispersant. This can be done by using a solution of the proper strength in your wash bottle and never using any water unless it has the right concentration of dispersant in it. You must always know exactly how much dispersant is in the water because it is an important factor in computation of the results. Keep in mind that you should always try several dispersants and use the one that works best; one of them may work on one sediment, and fail completely on another one. This is just a matter of trial and error.

2. Procedure for Dispersion. This technique is usable only if the sample is an unconsolidated or semi-consolidated sediment. Grain size analysis of hard shales is pointless as one can never break the material into individual grains; these can only be analyzed in thin-section. Ultrasonic treatment can disaggregate softer shales.

Place the material in a wide mouthed evaporating dish, add a small amount of water to which the proper concentration of dispersant has been added, and crush to pea-size lumps. Don a rubber glove and muddle the sample with the rubber-gloved finger until all the lumps, even the tiniest ones, have been crushed. Be careful not to lose any of the material that clings to the glove. Pour into a small bottle, again MAKING SURE THAT NONE IS LOST, screw a cap on the bottle, shake vigorously for a few minutes, and let stand over-night. If it is necessary to remove sand by wet-sieving, see page 20. Pour the suspension from the bottle into the dispersing cup (simply a drink mixer) and fill up about halfway with water which has the proper concentration of dispersant. Stir with the mixer for about 5 minutes. Rinse it out carefully into a liter cylinder, MAKING SURE THAT NONE IS LOST, and fill the cylinder up to exactly 1000 ml with water which has the proper concentration of dispersant. Now stir the material in the cylinder vigorously and let stand a day to check completeness of dispersion. Flocculation can be recognized by a curdling and rapid settling of clumps of particles, or by the presence of a thick soupy layer on the bottom of the cylinder passing abruptly into relatively clear water above. Or you may suck out a drop and place on a slide and examine with high power of petrographic microscope; if the particles appear as individuals and show Brownian movement, then dispersion is good; if they form clumps, chains or strings then dispersion is unsatisfactory.

To find the best dispersing agent, try several concentrations of several different dispersants with equal quantities of mud, and choose the one which makes the mud stand up longest or best without signs of flocculation. A frequent cause of flocculation is the use of too much sediment; try to use 15 gm or less.

IN ALL THESE PROCESSES, REMEMBER NEVER TO OVEN DRY THE MUD IF YOU EVER EXPECT TO MAKE A GRAIN SIZE ANALYSIS. HAVE YOU EVER TRIED TO DISAGGREGATE A BRICK OR A CHUNK OF ADOBE INTO ITS INDIVIDUAL GRAINS? THIS IS WHAT YOU'LL BE DOING IF YOU OVEN DRY THE MUD OR CLAY BEFORE YOU ANALYZE IT.

Separation of Sand From Mud

If the sample contains more than a few percent of material finer than 4ϕ, (0.0625 mm, the limit of silt and clay), it is usually necessary to separate the sediment into two fractions at this diameter value. The coarser material (sand) is then analyzed by sieving and the finer material (silt and clay) by pipette or hydrometer. The most critical part of the operation is getting the weight of each fraction. If nearly all the material finer than 4ϕ consists of silt, then the entire sediment may be dried, disaggregated and dry-sieved; then the material caught in the pan is weighed and pipetted directly.

If there is much clay-size material, however, dry-sieving will not work because the clay will gum up the screens; recourse must then be had to wet-sieving (some prefer to use this decantation though). Most operators use wet-sieving although it is admittedly slow and messy. It is probably the most accurate way.

1. Place the sample in a small bottle, fill approximately half full with water to which the proper concentration of dispersant has been added, and muddle with a rubber-gloved finger until all lumps are crushed (see under "2" on page 18). Screw the cap on the bottle and shake vigorously. Be sure not to lose any of the material.

2. Obtain a pan at least ten inches in width, and a wash bottle filled with no more than one-half liter of water to which the proper concentration of dispersant has been added (see pages 18, 19). (NOTE: If you are going to obtain the grain size distribution of the fines, you must use a bottle with dispersant added; if you are not going to pipette the fines, but simply weigh them to get the quantity present in the sediment, then use a bottle of distilled water because when you evaporate the fines to dryness you will be also weighing any dispersant you may have added). Now take the 62 micron (230 mesh) screen that is reserved for wet-sieving, and

dip it in distilled water or dispersant to get the mesh thoroughly wet on both sides, otherwise the sludge will not run through easily. Pour the sample from the bottle onto the screen while holding it over the large pan, and rinse out every grain from the bottle. Now rock the sediment in the screen back and forth while playing a fine jet of water on it, to wash the mud through the screen. Periodically wash down the under side of the screen with your water jet. Continue washing the sediment back and forth over the screen until the water runs through clear (this takes a little time). Use as little water as possible, because you should end up with a volume of water small enough to fit into the dispersing cup (page 20); by all means end up with less than a liter or else you will have too much to fit into the pipetting cylinder.

3. Pour the muddy water collected in the pan, into the dispersing cup, stir for 5 minutes, then pour into a liter cylinder and make it up to exactly 1000 ml. Continue as under "2" on page 19.

4 Take the sand remaining on the screen, dry it, disaggregate it, and sieve according to the standard method. Any material that passes the 4ϕ (230 mesh) screen on this sieving, retain and dump into the liter cylinder of mud, as it is material that should have gone through when you wet-sieved it.

5. Pipette the fine fraction according to instructions starting page 34.

In working with samples that contain more than a few percent of clay (i.e., those that have to be dispersed and then wet-sieved), it is impossible to get the total weight of "fines" (silt plus clay, material finer than 4ϕ or 0.0625 mm) directly; yet you must somehow obtain this figure in order to determine the percentage of fines in the sample. Here's the difficulty. Let us say you have a moist sample of sandy mud (or else a dry sample which you have to moisten in order to disperse it). It might appear easy to weigh the moist sample, then wet-sieve it and weigh the amount of sand retained on the sieve, which would be a certain percent of the total sample. But this is not true, since the total sample weight we determined was that of the sample plus a large amount of water. A second method would be to wet-sieve the sample and weigh the amount of sand; then evaporate the muddy water passed through the screen to dryness, and weigh that. This would be satisfactory were it not that further size analysis of the fine fraction is made impossible by over-drying. You can never analyze the fines once they have been baked.

There are three ways to solve this dilemma, and all involve an indirect determination of the mud content. In Method I (the pipette method), the sample is wet-sieved and the amount of sand weighed. The mud fraction, after dispersion, is placed in a cylinder and diluted up to exactly 1 liter, and stirred thoroughly. Twenty seconds after stirring, a pipette is inserted to 20cm depth and exactly 20 ml. of suspension withdrawn and put in a beaker. The suspension is evaporated, and the dried mud is weighed. The weight of mud multiplied by 50 gives the weight of mud in the total sample (since you have withdrawn exactly 1/50 of the muddy water in the cylinder). This assumes (1) that there has been no loss of fines in the various operations, (2) that during stirring all the particles have been perfectly uniformly suspended throughout the column of water, and that exactly 1/50 of a liter has been withdrawn, and (3) that there is negligible error in the weighing operation, especially in regard to absorption of moisture during weighing. Despite its drawbacks, this is probably the most accurate method and the one we will use. Results obtained by this method may be checked by using either of the two methods below.

Methods 2 and 3 split the sample into two portions with the mud-sand ratio being determined on one portion, and the analysis of the individual fractions being performed on the other portion. Both of these assume that both sub-samples have exactly equal proportion of constituents. This can be approached by very careful mixing. In Method 2 (wet-sieve replicate method) the sediment sample is spread out and thoroughly mixed. Then by repeated quartering or simly scooping portions from every part of the pile, one divides it into two subsamples. The first subsample is then wet-sieved, and the amount of sand weighed; the mud passing through is evaporated to dryness, allowed to come to equilibrium with room temperature and moisture content, and weighed. The percentage of sand versus mud is then computed for the first subsample and assumed to hold true for the second subsample. The second subsample is also wet-sieved, and the amount of sand weighed. Knowing the percentage of sand (from the first subsample), and the weight of sand in the second subsample, it is then possible to compute what the weight of mud in the second subsample should be. Then the mud in the second subsample can be dispersed and pipetted to obtain its size-distribution, because it has not been oven-dried.

If the sample contains over 40% mud, Method 3 (moisture-replicate method) may be used. Again the sediment is spread out and carefully divided into two subsamples. In this procedure the sediment is regarded as being made of three constituents: sand, mud, and water, and it is assumed that each subsample contains an equal proportion of these. Both samples are weighed moist. Then the first subsample is evaporated to dryness in an oven and weighed again. The loss in weight represents the water content, and is regarded as equalling the water content of the second subsample. The second subsample is then wet-sieved and the weight of sand obtained. Now, the moist total weight of the second subsample is known; the weight of sand in the second subsample is known; and the weight of water in the second subsample can be computed from the known percent of water in the first subsample. Subtraction thus gives the weight of mud in the second subsample, and the mud may then be pipetted to obtain its size-distribution. THIS PROCEDURE CANNOT BE USED IF THE SAMPLE CONTAINS LESS THAN 30-50% MUD, as the error is too great. The advantage of this method is speed.

These various methods may be summarized as follows:

I. The sample is dry

- A. It contains little or no clay: Dry sieve, weigh the sand fraction, weigh the pan fraction, and pipette the pan fraction.
- B. It contains considerable clay: Disperse and treat as a moist sample.

II. The sample is moist, or must be moistened to disperse it

- A. It contains little or no clay: Disperse, wet-sieve; use pipette or set-sieve replicate method to obtain percent of mud, then pipette the mud (or the sample may be air-dried, then treat as IA).
- B. It contains under 40% mud: Disperse, wet-sieve, use pipette or wet-sieve replicate method to obtain percent of mud; then pipette the mud.

C. It contains more than 40% mud: Disperse, wet-sieve, use pipette, wet-sieve replicate or moisture-replicate method to get percent of mud, then pipette the mud.

Decantation. Another way to get rid of small quantities of mud from the sediment sample is by decantation. Decantation is not recommended if the sediment contains more than 20-30% mud. Essentially, the sediment is stirred with water in a beaker, and after allowing a certain time for the coarser particles to settle to the bottom, the finer material remaining in suspension is siphoned off. This process is repeated until the water is clear. The advantage of this method is that it is somewhat easier than wet-sieving where only small quantities of mud are involved; also it can be used when no wet-sieve is available. The disadvantages are the tremendous volumes of water required and the necessity to evaporate all this water to dryness.

1. Decide what size material is to be decanted. If decantation is done in connection with sieving, then either 4ϕ or 4.5ϕ (62.5 or 44 microns) is chosen. Determine the settling time of particles of this size and density at the given temperature by means of Stokes' Law. For particles of 4.5ϕ diameter (44 microns), the settling velocity at 20°C is 0.174 cm/sec.

2. Set a 1000 ml beaker or cylinder on a small stand about 6" above the table. Fill it about one-third of the way with distilled water. Introduce the sample into the cylinder and stir vigorously, distributing all the sediment evenly throughout the column. Now, when the time comes for siphoning, you are to insert the sucking end of the siphon to a distance about one inch above the layer of sediment that has settled to the bottom of the cylinder. Measure the distance from the top of the water to the point where the siphon will be. Let us say that this is 50 cm. Then, particles of 4.5ϕ diameter will take 50/0.174 seconds, or 287 seconds, to settle that distance. In practice this means that at the end of 287 seconds all particles coarser than 4.5ϕ will have settled to the bottom and that only particles finer than this size remain suspended. After computing the desired time, stir vigorously and start timing the moment stirring stops. Fifteen seconds before withdrawal time, insert the siphon into the bottom of the cylinder, keeping the siphon 1" above the deposited sludge on the bottom. At the proper time, start siphoning the material off into an extra 1000 ml beaker. DO NOT SUCK ANY SEDIMENT OFF THE BOTTOM OF THE CYLINDER.

   Now, place the beaker in the oven and start evaporating it (this assumes that you are not going to analyze the size-distribution of the fines themselves, but only weigh them--remember never to oven dry the fines if you ever expect to obtain their grain size). Fill the cylinder with water, stir again, wait the same length of time, and siphon off another liter of muddy water into another beaker. Repeat this process until the water in the cylinder stands clear (you see this takes a great deal of water, and you cannot use this method for obtaining fines for pipetting because you end up with some 3 to 5 liters of water, and you can only have 1 liter for pipetting).

   Evaporate all the beakers to dryness, and add their weights (subtracting the weight of any dispersant you may have used). The sand in the bottom of the cylinder is also dried and weighed. Then the sand is sieved, and THE AMOUNT OF MATERIAL THAT PASSES THE FINEST SCREEN (say 4.5ϕ) MUST BE ADDED TO THE WEIGHT OF THE MUD IN THE EVAPORATED BEAKERS. The percentages of each can then be computed.

Grain Size Scales and Conversion Tables

The grade scale most commonly used for sediments is the Wentworth (1922) scale which is a logarithmic scale in that each grade limit is twice as large as the next smaller grade limit. The scale starting at 1mm and changing by a fixed ratio of 2 was introduced by J. A. Udden (1898), who also named the sand grades we use today. However, Udden drew the gravel/sand boundary at 1mm and used different terms in the gravel and mud divisions. For more detailed work, sieves have been constructed at intervals $^{2}2$ and $^{4}2$. The ϕ (phi) scale, devised by Krumbein, is a much more convenient way of presenting data than if the values are expressed in millimeters, and is used almost entirely in recent work.

U. S. Standard Sieve Mesh #		Millimeters (1 Kilometer)	Microns	Phi (ϕ)	Wentworth Size Class	
				-20		
		4096		-12		
		1024		-10	Boulder (-8 to -12ϕ)	
Use		256		-8		
wire		64		-6	Cobble (-6 to -8ϕ)	
squares		16		-4	Pebble (-2 to -6ϕ)	GRAVEL
5		4		-2		
6		3.36		-1.75		
7		2.83		-1.5	Granule	
8		2.38		-1.25		
10		2.00		-1.0		
12		1.68		-0.75		
14		1.41		-0.5	Very coarse sand	
16		1.19		-0.25		
18		1.00		0.0		
20		0.84		0.25		
25		0.71		0.5	Coarse sand	
30		0.59		0.75		
35	1/2	0.50	500	1.0		
40		0.42	420	1.25		
45		0.35	350	1.5	Medium sand	SAND
50		0.30	300	1.75		
60	1/4	0.25	250	2.0		
70		0.210	210	2.25		
80		0.177	177	2.5	Fine sand	
100		0.149	149	2.75		
120	1/8	0.125	125	3.0		
140		0.105	105	3.25		
170		0.088	88	3.5	Very fine sand	
200		0.074	74	3.75		
230	1/16	0.0625	62.5	4.0		
270		0.053	53	4.25		
325		0.044	44	4.5	Coarse silt	
		0.037	37	4.75		
	1/32	0.031	31	5.0		
Analyzed	1/64	0.0156	15.6	6.0	Medium silt	MUD
	1/128	0.0078	7.8	7.0	Fine silt	
by	1/256	0.0039	3.9	8.0	Very fine silt	
		0.0020	2.0	9.0		
Pipette		0.00098	0.98	10.0	Clay ↓	
		0.00049	0.49	11.0	(Some use 2ϕ or	
or		0.00024	0.24	12.0	9ϕ as the clay	
		0.00012	0.12	13.0	boundry)	
Hydrometer		0.00006	0.06	14.0		

PHI-MILLIMETER
CONVERSION
GRAPH

Given a ϕ diameter, to find the corresponding mm value:

(1) Select the diagonal line in which your diameter falls.

(2) Find ϕ diameter (on vertical scale).

(3) The mm equivalent is read on the horizontal scale. Use numbers at bottom of page to determine numerical values, and the mm values written on the lines will indicate the decimal point.

Example: 2.6ϕ = 0.165 mm
11 mm = -3.45ϕ

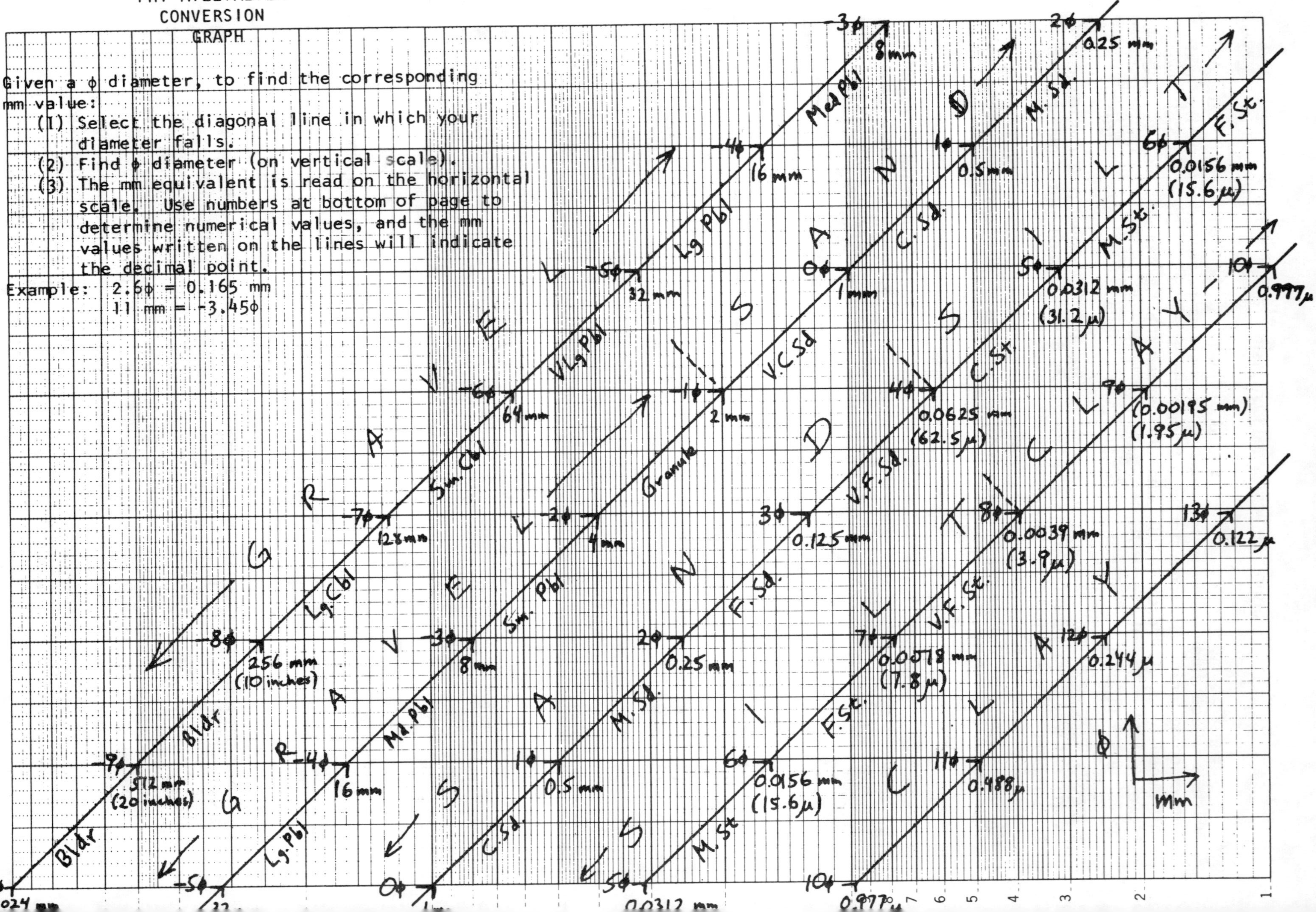

Grain Size Nomenclature

The basis of the classification is a triangular diagram on which are plotted the proportions of gravel (material coarser than 2mm), sand (material between 0.0624 and 2mm), and mud (defined as all material finer than 0.0625mm, i.e., silt plus clay), as shown in the triangular diagram. Depending on the relative proportions of these three constituents, fifteen major textural groups are defined--for example, sandy conglomerate, slightly conglomeratic mudstone, or sandstone. This classification is presented in detail by Folk (Jour. Geol. 1954); and also Folk, Andrews & Lewis (NwZd).

To place a specimen in one of the fifteen major groups, only two properties need to be determined: (1) how much gravel (material coarser than 2mm.) it contains--boundaries at 80, 30, 5 per cent, and a trace; and (2) the ratio of sand to mud (silt plus clay) with boundaries at 9:1, 1:1, and 1:9.

The proportion of gravel is in part a function of the highest current velocity at the time of deposition, together with the maximum grain size of the detritus that is available; hence even a minute amount of gravel is highly significant. For this reason the gravel content is given major emphasis, and it is the first thing to determine in describing the specimen. This is best done on the outcrop by naked-eye examination, perhaps aided by a percentage comparison chart; thin sections and hand specimens commonly give too small a sample to be representative of the gravel content. Using this scheme, a specimen containing more than 80 per cent gravel is termed "conglomerate"; from 30 to 80 per cent gravel, "sandy conglomerate" or "muddy conglomerate"; from 5 to 30 per cent gravel, "conglomeratic sandstone" or "conglomeratic mudstone"; from a trace (say 0.01 per cent) up to 5 per cent gravel, "slightly conglomeratic sandstone" or "slightly conglomeratic mudstone"; and a specimen containing no gravel at all may range from sandstone through mudstone, depending on the sand:mud ratio.

The proportion of sand to mud is the next property to be determined, reflecting the amount of winnowing at the site of deposition. Four ranks are defined on the basis of the sand:mud ratio; in the nonconglomeratic tier, these are sandstone (ratio of sand to mud over 9:1), muddy sandstone (ratio 1:1 to 9:1), sandy mudstone (ratio 1:9 to 1:1), and finally, mudstone (ratio under 1:9). The ratio lines remain at the same value throughout the triangle: e.g. sandy conglomerate is also divided from muddy sandy conglomerate by a sand:mud ratio of 9:1. This division is fairly easy to make with a hand lens, unless a large amount of coarse silt and very fine sand is present.

These two simple determinations are sufficient to place a specimen in one of the fifteen major textural groups shown in Table 1. One might simply stop at this point and say no more about the grain size; yet a great deal of information is gained by specifying, whenever practicable, the median diameter of each of the fractions present. Thus two specimens belonging to the conglomeratic sandstone group have quite different significance if one is a bouldery fine sandstone and the other is a pebbly very coarse sandstone. The detailed breakdown will be used in all our class work.

These fine subdivisions are determined by specifying the median diameter of each fraction considered independently of any other fraction that may be present. For some specimens this is, of course, not possible; but in most it can be accomplished sufficiently well for field purposes, especially if the material is bimodal. The size terms of Wentworth (1922) are used for the various classes. Thus, if gravel is present, one determines whether the median of the gravel fraction considered alone falls in the granule, pebble, cobble, or boulder class; for example, the major group of sandy

G, gravel; g, gravelly;
(g), slightly gravelly
S, sand; s, sandy
M, mud; m, muddy

Specify median size of gravel wherever present.
Specify median size of sand only in the stippled area.
Specify composition of mud (whether silty, muddy, or clayey) wherever present.

Examples:
sG, sandy cobble gravel
(g)sM, slightly granular fine sandy silt
mS, clayey medium sand

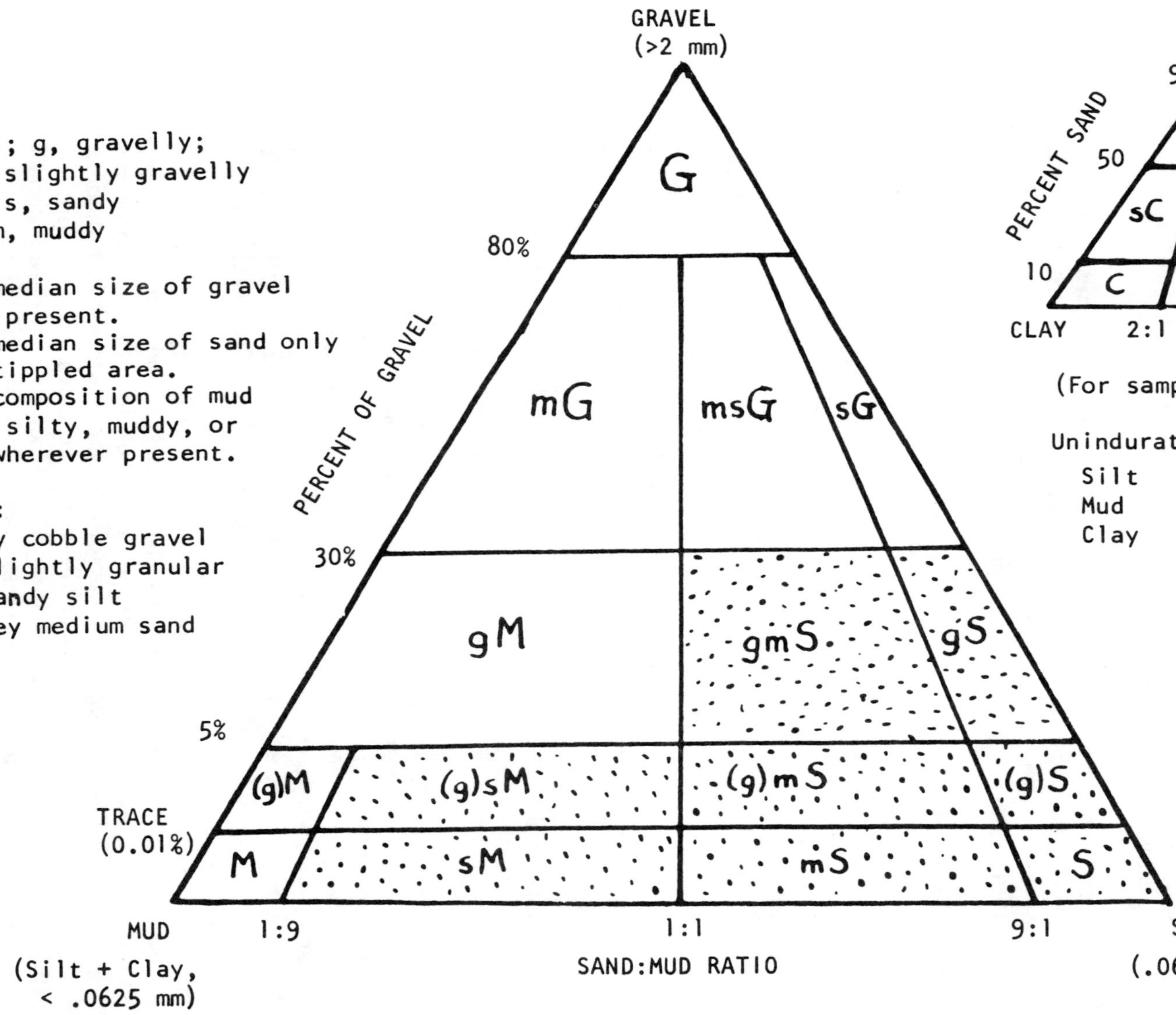

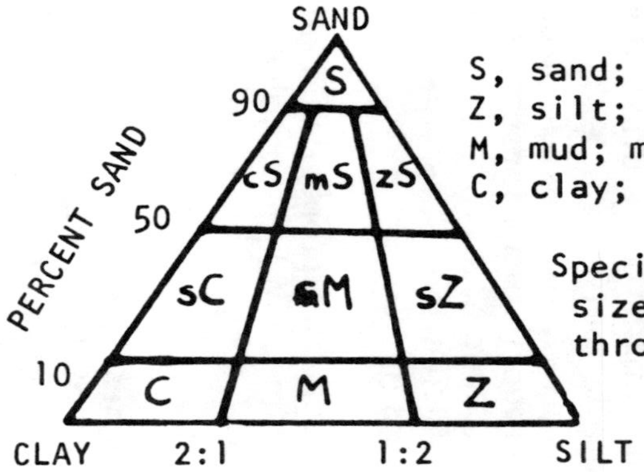

S, sand; s, sandy
Z, silt; z, silty
M, mud; m, muddy
C, clay; c, clayey

Specify median size of sand throughout.

(For samples lacking Gravel)

Unindurated	Indurated	Fissile
Silt	Siltstone	Silt-shale
Mud	Mudstone	Mud-shale
Clay	Claystone	Clay-shale

For details, see Folk, Jour. Geol. vol. 62, p. 345-351, July 1954.

TABLE I*

Terms Applied to Mixtures of Gravel, Sand, and Mud
Delimited in the triangular diagram page

	Major Textural Class	Examples of Usage
G.	Gravel	Cobble gravel
	Conglomerate	Granule conglomerate
sG.	Sandy gravel	Sandy pebble gravel
	Sandy conglomerate	Sandy boulder conglomerate
msG.	Muddy sandy gravel	Muddy sandy granule gravel
	Muddy sandy conglomerate	Clayey sandy pebble conglomerate
mG.	Muddy gravel	Silty boulder gravel
	Muddy conglomerate	Muddy pebble conglomerate
gS.	Gravelly sand	Pebbly coarse sand
	Conglomeratic sandstone	Granular very fine sandstone
gmS.	Gravelly muddy sand	Pebbly silty fine sand
	Conglomeratic muddy sandstone	Bouldery muddy coarse sandstone
gM.	Gravelly mud	Cobbly clay
	Conglomeratic mudstone	Pebbly siltstone
(g)S.	Slightly gravelly sand	Slightly granular medium sand
	Slightly conglomeratic sandstone	Slightly pebbly coarse sandstone
(g)mS.	Slightly gravelly muddy sand	Slightly pebbly muddy medium sand
	Slightly conglomeratic muddy sandstone	Slightly cobbly silty fine sandstone
(g)sM.	Slightly gravelly sandy mud	Slightly granular fine sandy mud
	Slightly conglomeratic sandy mudstone	Slightly pebbly coarse sandy claystone
(g)M.	Slightly gravelly mud	Slightly pebbly clay
	Slightly conglomeratic mudstone	Slightly cobbly mudstone
S.	Sand (specify sorting)	Well-sorted fine sand
	Sandstone (specify sorting)	Poorly sorted medium sandstone
mS.	Muddy Sand	Well-sorted silty very fine sand
	Muddy sandstone	Muddy coarse sandstone
sM.	Sandy mud	Fine sandy clay
	Sandy mudstone (specify structure)	Coarse sandy siltstone (if fissile, coarse sandy silt-shale)
M.	Mud	Silt
	Mudstone (specify structure)	Mudstone (if fissile, mud-shape)

*Both unconsolidated and consolidated equivalents are shown in this table. The underlined terms are further specified as to their grain size, as shown in the examples.

conglomerate can be subdivided into sandy granule conglomerate, sandy pebble conglomerate, sandy cobble conglomerate, adn sandy boulder conglomerate. For the sand fraction, the median is also estimated separately, using the standard Wentworth grades of very coarse sand, coarse sand, medium sand, fine sand, and very fine sand. This can be done very easily by reference to a comparison set of sand grains of the several sizes.

For muds a somewhat different procedure is used, the name depending on the relative proportion of silt versus clay. This proportion is usually very difficult to determine with a hand lens, and the only really satisfactory way is to make a thin-seciton (preferable) or a grain-size analysis by pipette or hydrometer. In many samples it might be best just to use the broad term "mud" and not attempt to split it any further. But the mud fraction of many sediments is obviously composed dominantly of silt, while the mud fraction of others is just as certainly composed largely of clay; therefore, it is considered worthwhile to make an attempt, if at all practicable, to estimate this ratio. A threefold division is suggested: if the mud fraction contains more than 67 per cent silt (i.e., silt-to clay ratio greater than 2:1), the material should be called "silt" or silty"; if more than 67 per cent clay is present, it should be called "clay" or "clayey" and for intermediate mixtures, the term "mud" or "muddy" (used in a restricted sense) is proposed. Thus the major group of muddy sandstone may be divided into clayey fine sandstone, silty very fine sandstone, muddy coarse sandstone, and so on, since both the grain size of the sand fraction and the mud composition are to be specified.

The complete series of major textural groups is presented in the table. Both unconsolidated and consolidated equivalents are given for each group. At the right of the group name is given an example of how the terms are to be further specified, depending on the median grain size of each fraction present. Sand, silty sand, and slightly conglomeratic sand may be further described by mentioning the degree of sorting.

The major textural groups of "mudstone" and "sandy mudstone" should be modified according to their structure, specimens with a well-developed closely spaced parting parallel with the beds being termed "shale," regardless of whether they are composed of clay, silt, sandy clay, mud, or any other mixture of materials. This additional modification is presented in table 2.

TABLE 2

Classification of Fine-Grained Rocks Based on Grain Size, Induration, and Structure

Textural Class	Unindurated	Indurated, Not Fissile	Indurated and Fissile
Z	Silt (> 67 per cent silt)	Siltstone	Silt-shale
M	Mud (intermediate)	Mudstone	Mud-shale
C	Clay (> 67 per cent clay)	Claystone	Clay-shale
sZ	Sandy silt	Sandy siltstone	Sandy silt-shale
sM	Sandy mud	Sandy mudstone	Sandy mud-shale
sC	Sandy clay	Sandy claystone	Sandy clay-shale

Suggested Outline for Detailed Study of Texture

In routine field examination the determination of a grain-size name is probably sufficient; but for detailed work the following model for textural description is offered. Following the specimen number, stratigraphic and geographic position, and other pertinent information, the method of study is recorded, whether it be by hand-lens, binocular-microscope, thin-section, or mechanical analysis. A space is left for the grain-size name, which, of course, is filled in after the description has been made. Next, the size characteristics of the sediment as a whole and then of each of the three size fractions are reported. For the entire sediment, the range between the sixteenth and eighty-fourth percentiles of the size distribution should be given, as this gives a much truer picture of the nature of the sediment than simply mentioning the extreme range. Possible discontinuities in the distribution (bimodality) are very important and should be described in detail. Following this information appears a discussion of the relation of the three textural end-members of one another, including the following characteristics: (1) homogeneity (Does the rock have coarse and fine layers, or are the size fractions thoroughly intermixed?); (2) thickness and type of bedding (is it massive, random, or cyclic parallel, hummocky, graded-bedded, or cross-bedded? Is the bedding caused by a physical parting or a change in composition, texture, or color?); (3) orientation of grains; and, finally, (4) particle shapes. Mention is then made of the correlation of size with composition (for example, the gravel fraction might consist entirely of shell fragments, whereas the sand fraction was entirely quartz). An example of this method follows:

I. Sample number and location: (Give pertinent date)

II. Method of study: Hand lens, in outcrop

III. Grain-size name: Pebbly muddy fine sandstone

IV. Detailed size description

- A. Entire sediment
    1. Median: 0.2-0.3 mm.
    2. Sorting: Poor
    3. 16-84 per cent range: Clay size to about 5 mm.
    4. Extreme range: Clay size to 20 mm.
- B. Gravel fraction
    1. Per cent: 15
    2. Median: About 6 mm.
    3. Range: 3-20 mm., in itself poorly sorted
- C. Sand fraction
    1. Per cent: 55-65
    2. Median: Probably in fine-sand class, about 0.2 mm.
    3. Range: Up to 0.5 mm., rather well sorted
- D. Mud fraction
    1. Per cent: About 20-30
    2. Silt versus clay: About half clay(?)
    3. Median of silt portion: Not determined in the field

E. Chemical precipitates
   1. Per cent: Probably under 5
   2. Type: Mostly calcite, a little quartz

V. Textural relations

A. Gravel tending to be segregated in vague bands parallel with the bedding, whereas sand and mud are intimately intermixed; bedding 3 feet thick, random parallel to hummocky, a response to grain-size variation; gravel shows weak imbrication, dip southeast; pebbles subround, sand grains subangular

B. Distribution strongly bimodal, with little material between 0.5 and 3 mm.; separate modes in pebble and fine sand size

C. Gravel fraction composed entirely of chert, dull white; sand is chiefly quartz; mud not identifiable in the field

Every terminology has its weak points, and this one is certainly no exception. For example, a unimodal specimen consisting of a well-sorted mixture of coarse silt and very fine sand should, according to the ternary diagram, be placed in one of the fifteen textural groups by determining the ratio of sand to mud (the latter in this case consisting entirely of silt, with no clay); yet this proportion would be exceedingly difficult to determine, and the rock might be termed anything from a "well-sorted siltstone" to a "well-sorted silty very fine sandstone." The difficulty in placing such boundary specimens is present in every classification but should not be considered a serious handicap.

It may be argued that by far the greatest bulk of sediments and sedimentary rocks lies in the bottom, nongravelly tier of the diagram, which includes only four major classes; nevertheless, by virtue of the fact that the grain size of the sand fraction (five grades) and the silt:clay ratio (three divisions) must be determined, a very large number of subclasses is available (fig. 1b)--and this does not even include the mention of sorting within groups and fissility, which should be apart of the rock name. Thus the grain-size classification is capable of considerable "lumping," if one wishes to confine himself to major group names, or to a much more intricate subdivision, if one wishes to study a series of rocks that are closely related in grain-size distribution.

Making Distribution Maps Showing Grain Size of Sediments

A very useful way to summarize data from recent sediments studies that involve sampling large areas (bays, offshore areas, deltas, etc.) is to make a map of the geographic distribution of grain size classes. This can be done very easily using the grain size triangle nomenclature for sand-silt-clay mixtures. One takes his sampling points and records for each one the percent of sand in the sediment. This data is contoured using values of 10%, 50%, and 90% sand as contour lines (these are the boundary lines of the four "sandiness" classes of the grain size triangle). Next, one records at each sampling point the percent silt in the mud fraction (i.e. percent silt divided by percent silt plus clay). These values are contoured at the 33% and 67% lines. The map is now covered with two independent, intersection contour systems, and each "cell" bounded by these contours is a grain size class (e.g. the sector bounded by the 50 and 90% sandiness contours and the 33 and 67% siltiness contours is termed "muddy sand"). These different cells may not be labeled, colored or shaded to bring out the areal distribution of the different grain size classes. If the area has gravel, gravel porportions can be contoured in a similar fashion.

Size Analysis by Sieving

It is assumed that the sand has already been disaggregated and that clay or mud, if present in considerable amounts, have been removed. These methods have been described in previous sections.

1. Using a sample splitter, obtain about 30 to 70 grams of sample. If there are numerous screens to be used in the analysis, use the larger weight; if only 4 to 6 screens, use the smaller weight (this is to prevent clogging the screens with too much sand, a frequent cause of poor results). If there is <u>any</u> gravel (material coarser than 2 mm), see note at end. Spread the sand evenly along the hopper and <u>stir</u> it well to mix it thoroughly (do not shake as this will make the finer grains travel to one end).

2. Weigh the split sample to 0.01 gram.

3. Select the screens to be used. For accurate work, use the ¼ ϕ set; for rough work, use the ½ ϕ set. If you are doing any research, it is senseless to use the 1ϕ interval as this is too broad an interval to get useful data. The student sets are 1ϕ, but these are just for practice. Clean the screens thoroughly using the procedure below. This is <u>especially</u> important if you are going to make mineral studies on the samples after screening.

4. Nest the screens in order, coarsest at the top, pan on the bottom. If the stack is too big to fit into the Ro-Tap, it will have to be sieved in several stacks, starting with the coarsest sizes. Pour the sample in to the top sieve and shake gently by hand. Remove all the screens that are too coarse to catch any grains. Place cover on the stack.

5. Place the screens in the Ro-Tap, FASTEN VERY TIGHTLY, and sieve for 15 minutes. For all scientific work, the Ro-Tap machine must be used, 8 inch diameter screens must be used, and the sieving time should be constant for all samples (15 minutes is the accepted time). For the small 3" student sets, sieving must be done by hand, using a rotary motion with a bump.

   If the analysis must be sieved in two stacks, remove the first stack from the Ro-Tap; take the material caught in the pan, and dump it carefully into the top of the second stack (be sure there is another pan on the bottom of the second stack). Place the first stack on its now empty pan again.

6. Take a large sheet of brown paper (at least 18" X 18"), crease it in the middle, and lay it on the table. Then take a sheet of glazed notebook paper (or larger), crease it and lay it in the center of the large sheet. Now hold the coarsest screen over the small sheet of paper and carefuly pour out the sand. Then invert the screen and tap it gently with the heel of the hand. YOU MUST TAP IT DIAGONAL TO THE MESH OR THE SCREEN WILL BE DAMAGED.

7. On a spare piece of glazed, creased paper, place the balance pan. Carefully pour the sand from the pile on your two papers, into the balance pan.

8. Replace the two creased papers as before and now hold the sieve upside down and pound it sharply on the paper, STRIKING THE TABLE EVENLY WITH THE RIM (otherwise you will dent the screen). Add the sand thus jarred loose to the balance pan. Make sure that all grains end up inside the balance pan by repeating this process. Weigh the sample in the balance pan to 0.01 gm. (if you have less than 1.0 grams in any sieve fraction, it should be weighed on the chemical balance to 0.001).

9. Examine each sieve fraction (after it is weighed) under the binocular microscope and estimate the percent of aggregates in each fraction. THIS IS ABSOLUTELY NECESSARY FOR VALID WORK! The best way to do this is by spreading grains on a micropaleo grid and counting 100 of them. This takes only a very little time. Start estimating at the coarsest sizes and work down until NO MORE aggregates appear. If any fraction over 25% aggregates, it should be re-disaggregated and run through the screens again. Record the percentage of aggregates in each size--these must be subtracted from the sample weight as shown in the sample computation.

10. Store the sample in corked vials, paper envelopes, or folded paper packets. Label each fraction as to both coarser and finer limits, and sample number.

DATA FORM

Held on: Mesh	Mm	ϕ	Raw Weight	% Aggre- gates	Corrected Weight	Cumulative Weight	Cumulative Pct.	Indiv. Pct.
18	1.0	0.0	3.5 gm	20	2.8	2.8	9.1	9.1
35	0.5	1.0	10.7	10	9.6	12.4	40.3	31.2
60	0.25	2.0	8.3	5	7.9	20.3	65.9	25.6
80	0.19	2.5	5.6	-	5.6	25.9	84.0	18.1
120	0.12	3.0	2.9	-	2.9	28.8	93.5	9.5
230	0.06	4.0	1.3		1.3	30.1	97.7	4.2
Pan	—		0.7		0.7	30.8	100.0	2.3

To get cumulative percent, divide each figure by the last number in the Cumulative Weight column. Do not obtain cumulative percent by adding up individual percentages (last column) as "rounding-off" error will distort the cumulative percentages.

This analysis illustrates another trick: if you are sieving at a coarse screen interval, it is wise to insert an additional screen at about the modal diameter (the most abundant grade size). This makes curve plotting much easier.

IF the sample contains any gravel (Material coarser than 2.0 mm):

To get a representative sample of the gravel fraction, it is necessary to have a much larger sample, say 0.5 to 2 or even more kilograms. In this case, place the total sample on the 2 mm. screen and sieve by hand. Weigh the total amount of gravel retained, and the total amount of sand passing through. This enables you to obtain the percentage of gravel in the sample.

Sieve the entire amount of gravel by hand through as coarse screens as are available. For pebbles too large, a set of wire squares of proper diameter is constructed, and the pebbles passed through endwise and counted. The number of pebbles unable to pass through each screen is then weighed. Screen and weigh all the pebbles in this manner.

Split the total amount of sand to 30 to 70 grams and sieve as customary. Now, we have sieved only a small fraction of the total amount of sand in the sample. To compute the analysis, it is necessary then to multiply each sieved sand fraction by a "splitting factor", which is (1) the total weight of sand in the sample (obtained in the first operation where gravel was separated out) divided by (2) the weight of the sand split that was actually sieved.

Say we start with a sample of 1500 grams, and 600 grams is gravel and 900 grams is sand. Then we screen all of the gravel fraction, and a 105-gram split of the sand fraction. The splitting factor is then 900/105, or 8.571. Each sand fraction is multiplied by 8.571 to give the weight of that fraction in the entire original sample. These weights are then cumulated with the gravel weights and percentages computed as usual.

	Held on:	Wt.	X Splitting Factor (8.571)	Cum. Wt.	Cum. %
Gravel:	16 mm	150	-	150	10.0%
	8 mm	80	-	230	15.3%
	4 mm	140	-	370	24.7%
	2 mm	230	-	600	40.0%
Sand:	1 mm	52	446	1046	69.8%
	0.5 mm	30	257	1303	87.0%
	0.25 mm	18	150	1458	97.3%
	0.125 mm	5	42	1500	100.0%

Size Analysis by Settling Tube

For rapid, generally less accurate work, grain size of sand may be measured by settling the grains through a water column. Fancy devices are available to make sure all the grains start settling together. Best results are obtained with water columns at least 10 cm. in diameter, and with a small number of grains. Results are recorded by continuous weighing of grains accumulating at the bottom, or by automatic recording of pressure differences. Both "sievers" and "settlers" are strongly opinionated as to which is the best method of measuring grain size--one may just as well ask, "Which is the best way of doing carpentry--a saw or a hammer?"

Pipette Analysis by Silt and Clay

There are several methods of analysis of silts and clays (finer than 4ϕ or 62 microns diameter) in common use. The most common method is by pipette. Hydrometer and decantation methods are more difficult and less accurate, and many electronic sediment-analysis devices are now available (and expensive). All are based on the settling velocity of the particles, usually computed on the bases of Stokes' law: in the pipette method, one sucks off a small volume of suspension, evaporates and weighs the residue remaining suspended at the given time; in the hydrometer method one measures the density of the suspension, which depends on the grains still suspended after a given time, dries and weighs them. Electronic devices measure the density of the suspension. No settling method is very apropos for particles finer than 11ϕ or 12ϕ, where Brownian motion interferes with settling.

For pipette analysis, all material coarser than 4ϕ (62 microns) must be removed from the sample, either by wet or dry sieving; the best amount of sample to work with is approximately 15 grams, although analyses can be run on samples as small as 5-10 grams. With more sample the grains interfere wth each other too much in settling and may flocculate; with less sample, the experimental error in weighing becomes large with respect to the sample.

It is assumed that you start with a sample already dispersed (see page 22). Pour the dispersed sample in the liter cylinder, add dispersant and water to bring it up exactly to 1000 ml. Stir vigorously and let the cylinder stand for a day; if it shows no signs of flocculation, it is ready for analysis; if it starts to flocculate (see page 24) dispersion will have to be redone or the sample discarded.

Prepare a data sheet giving diameters, depths and times of withdrawal listed on it. Obtain an accurate timer (watch with second hand will do), a stirring rod, pipette, and an array of 50 ml beakers.

Stir the cylinder vigorously, starting at the bottom and working up until all the material (even the coarsest) is distributed uniformly throughout the column. End up with long, smooth strokes the full length of the column, from the very bottom until the rod breaks surface. THIS IS EXCEEDINGLY IMPORTANT. As soon as the stirring rod emerges for the last time, start the time. At the end of 20 seconds, insert the pipette to a depth of 20 cm and withdraw exactly 20 ml. This is the most important single step, as on it the subsequent analyses are based, so make it as close to 20.0 ml as you possibly can.

Continue withdrawals at the specified time intervals. It is unnecessary to stir the suspension at any time after the first stirring; however if you need to you can stir over again and start timing from the new stirring; for example if you forget to take the 2-hour sample, you can stir the cylinder again and wait another two hours. In general this is not desirable and should be avoided if possible.

Years of experience have shown that this is the best procedure for making withdrawals: fifteen seconds before the time of withdrawal, grasp the rubber tube of the pipette with the left hand and convey it to the mouth. With the right hand, insert the glass part of the pipette to the proper depth, and steady it by resting the right hand on the rim of the cylinder. With your left hand grasp the pipette at the place where the rubber tube is attached to the glass, so you will be able to get an exact volume in the pipette by using your left index finger to clamp the rubber tube shut against the end of the glass tube. When the time arrives, begin sucking very rapidly and draw the suspension up about an inch above the volumetric mark, and clamp with the left finger. Then let the suspension down very slowly by slowly unclamping the finger, until the liquid surface hits the volumetric mark. If you are off by no more than 1/8" either way it is sufficiently accurate. Remove the pipette, expel the suspension into a weighed 50 ml. beaker, suck up 20 ml of distilled water to rinse out the pipette, and expel the rinse water into the same beaker.

Place the beakers in an oven with cover glass on top, and evaporate them to dryness (they should be dried at least 25 hours at 100-130° C). Remove them from the oven and let them cool to room temperature. Let them remain open in the room where weighing is to take place for several hours, until they come to equilibrium with the moisture content of the atmosphere (the thin clay films in the beakers will begin to absorb water very rapidly when they first come out of the oven, consequently gain rapidly in weight at first). Weigh them to 0.001 gm. on a chemical balance; if possible weigh all beakers at nearly the same time.

NOTE: Some very careful analysts bake their clean beakers, keep them at absolute dryness in a desiccator, and maintain a desiccating chemical (calcium chloride) in the scales to keep everything as dry as possible. If it were possible to keep everything perfectly dry, this would be the best method; but in actual fact there is probably more error in this procedure because of the initial rapid weight gain of the samples when exposed, even for a few seconds, to ordinary room air. Obviously you cannot weigh all the samples with the same speed, therefore some will gain more moisture than others so you will never be sure of the real weights. On the other hand, if you treat them all equally by letting them come to equilibrium with the room atmosphere, you will get more consistent results. After all, in the natural sediment there is considerable moisture absorbed in the clay particles. But remember: if you weigh each beaker in the "room-moisture" state, you must do all your other weighings in the same way.

Computation. Tabulate the data as shown on the next page. Subtract the weight of the cleaned, air-dried beaker from the weight of the beaker plus sample. Now, figure the weight of dispersant in the entire liter of water (computed by using molecular weight and normality), and divide this value by 50; this gives the weight of dispersant in each of the small beakers. This weight must be subtracted from the weight of the sample.

The principle behind the computation is this: if the fine sediment is uniformly distributed throughout the entire 1000 ml. column by stirring, and we draw off exactly 20 ml. at stated times, then the amount of mud in each withdrawal is equal to 1/50 of the total amount of mud remaining suspended in the column at that given time and at that given depth (i.e., the amount of mud finer than the given diameter; all particles coarser than the given diameter will have settled past the point of withdrawal). The first withdrawal is made so quickly after stirring (20 seconds) and at such a depth that particles of all sizes are present in suspension; therefore if we multiply the weight of this first withdrawal by 50 (after subtracting dispersant weight), we will obtain the weight of the entire amount of mud in the cylinder. Then if we withdraw a sample at a settling time corresponding to a diameter of 6ϕ, and multiply it by 50, then we know that the product represents the number of grams of mud still in suspension at this new time, therefore the grams of mud finer than 6ϕ. Similarly we can compute the number of grams present at any size, and obtain the cumulative percentages as shown below:

First, we wet-sieved the sample. Let the weight of sand caught on the 4ϕ (62 micron) screen be called S. Then, 50 times the first pipette sample (the 20-second sample) equals the total amount of mud in the cylinder, hence the amount of fines passing through the screen; call this amount F. The percentage of sand in the sample is then 100S/(S+F). Let us now denote by "P" the quantity obtained by multiplying each later pipette sample by 50; then to obtain the cumulative percentages of the total sample directly from the pipetting data, we substitute the proper values in the equation,

$$\text{CUMULATIVE PERCENT COARSER} = \frac{100\,(S + F - P)}{S + F}$$

and these values may be plotted directly on the cumulative curve.

An example of the computation is given on the following page.

Graphing. Results for gravel (if any), sand, and mud should be combined in a smooth, continuous cumulative curve. Normally, the analysis stops at about 10ϕ, and there still may be a large amount of sediment finer than that. To obtain grain-size parameters (mean, σ_I, etc.) the cumulative curve is extended in a straight line on ordinary arithmetic (squared) graph paper from the last date point (usually 10ϕ) to 14ϕ at 100 percent. This assumed that essentially all clay particles are larger than 14ϕ (0.06 micron), and that the clay mode is somewhere near 12ϕ (0.24 micron). Grain size data may then be obtained from this extrapolated curve.

ϕ diam. csr. 4	Withdrawal Depth cm.	Time at 20°C	Weight spl. and beaker	Weight beaker	Wt. spl.	Less Dispers. .037 g.	X50	Cumulative Percent (see formula above)
	--weight obtained by wet-sieving--				32.53(S)			76.7(100S/S+F)
4	20	20s	10.553	10.321	.232	.195	9.75(F)	76.7
5	10	1m56s	11.631	11.486	.145	.108	5.40(P)	87.1
6	10	7m44s	9.903	9.794	.110	.073	3.64(P)	91.2
7	10	31m	10.608	10.510	.098	.061	3.04(P)	92.6
8	10	2h03m	12.781	12.718	.063	.025	1.25(P)	96.9
9	5	4h06m	9.726	9.672	.054	.017	0.85(P)	98.0
10	5	16h24m	10.301	10.256	.045	.008	0.40(P)	99.0

NOTE two things: that if you keep the withdrawal depth constant, for each ϕ diameter smaller you have to wait four times as long; but if you withdraw from half the depth, you cut the waiting time in half. This is because the settling velocity is proportional to the square of the particle diameter. You need not stick religiously to the times above; you can always make withdrawals at odd times and compute the diameter for that time. In general you should take samples at ½ϕ intervals from 4 to 6 ϕ, and at full ϕ intervals from then on.

The times and depths of withdrawals for particles of any size are given by:

$$T_{min} = \frac{\text{Depth in cm.}}{1500 \cdot A \cdot d^2 \text{ (mm)}}$$

in which T is the time in minutes, d^2 is the square of the particle diameter in mm, and A is a constant which depends upon viscosity of the water (a function of temperature), the force of gravitation, and the density of the particles. The table giving values of A for various temperatures and particle densities is given below.

A values

DENSITY

Temp. °C	Clays, Quartz 2.65	3.00	Amphiboles 3.35
16	3.23	3.92	4.60
20	3.57	4.33	5.08
24	3.93	4.76	5.60
28	4.30	5.21	6.21
32	4.68	5.67	6.60

For quartz or clay minerals at 24°C, these times work out as follows:

Depth cm	ϕ	Mm	Times of withdrawal	
20	4.0	.0625	20 seconds	
20	4.5	.044	1m45s	← restir
10	5.0	.031	1m45s	These times counted from the restirring
10	5.5	.0221	3m28s	
10	6.0	.0156	6m58s	
10	7.0	.0078	28m	
10	8.0	.0039	1h51m	
10	9.0	.0020	7h24m	
5	10.0	.00098	14h50m	
5	11.0	.00049	59h20m	
5	12.0	.00024	237h20m	
5	13.0	.00012	949h	

(If you use Waddell's sedimentation formula, multiply all the "A" values by 0.64 and use this instead).

Graphic Presentation of Size Data

The data obtained from grain-size analysis may be plotted in many different ways. For some purposes, one method is best; for other purposes, other methods are best. Get familiar with all methods, so that you do not become so blindly used to using one method that you ignore other ways which may be more suitable for a given problem.

All the methods use grain size as the abscissa (horizontal scale) and some measure of percentage frequency as the ordinate (vertical scale). Grain size analyses may either be plotted directly in millimeters, using a logarithmic-base paper; or they may be plotted in phi units (ϕ), in which case arithmeticbase paper is used. The latter is much more convenient and accurate to read.

1. Histogram. A histogram is essentially a bar graph in which the percentages for each grade size are plotted as a column. It is very easy to prepare and one can easily interpret general features of the sediment. However, it is a pictorial method, no more, and cannot be used for determination of any statistical parameters such as median, sorting, etc. Furthermore its shape is greatly affected by the sieve interval chosen; the same sample may look entirely different if it is analyzed on a different set of screens and another histogram prepared. Nevertheless it proves of value in plotting distribution of sediments on a map or stratigraphic section, as the heights of the columns may be more easily compared by eye than if the data were plotted as cumulative curves. For pictures, yes; otherwise, no.

2. Cumulative Curve, Arithmetic Ordinate. This is the most commonly used method. As the abscissa, one may use either millimeters (in which case he must use semi-log paper) or phi units (ordinary "squared" arithmetic paper). The ordinate is an arithmetic scale running from 0 to 100%; grain size is plotted on the abscissa with coarser particles to the left (this is customary in all size analysis plots). Cumulative percentages of the sediment are plotted on this graph; for example, if 32% of the material is coarser than 2ϕ (caught on the 2ϕ screen) then 32 is plotted as the ordinate against 2.0 as abscissa. Draw a curve through all the resulting points. YOUR CURVE MUST PASS THROUGH ALL THE PLOTTED POINTS -- NEVER USE A FRENCH CURVE TO "SMOOTH" OUT THE GRAPH. The sample analysis normally forms an S-shaped curve. The advantage of this curve is that all statistical parameters may be read from it exactly, thus one can compare samples quantitatively as to median, skewness, etc. The shape of the curve is independent of the sieves used. Its only disadvantage is that it is difficult for the untrained eye to look at the curve and interpret it at a glance; it is not "pictorial". Also if the sieve interval is wide, sketching the curve between data points is subject to considerable error.

3. Cumulative Curve, Probability Ordinate. Most sediments tend to approach the "normal probability curve" in their size frequency distribution--in other words, most of the particles are clustered about a given size, with less and less material on each side of this size. If the cumulative curve of a sediment following the normal, symmetrical probability distribution is plotted on probability paper, the result is a perfectly straight line whose position depends on the average particle size and whose slope depends on the sorting. This happens because the probability scale is very condensed in the middle of the scale (30 to 70%) and very much expanded at the ends (under 10 or over 90%), thereby straightening out the S-shaped curve which would result if arithmetic ordinates were used. Thus it is very valuable for studying the departure of sediments from the normal probability law. Moreover, since the "tails" are straightened out and the sample tends to plot as a straight line, it is possible to read the statistical parameters with much greater accuracy because of the ease of interpolation and

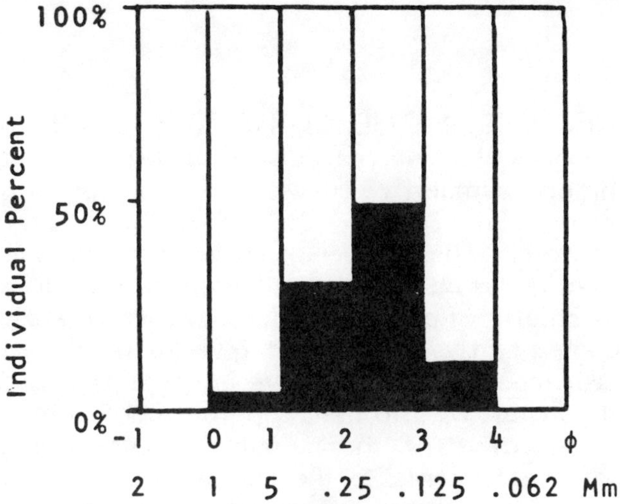

Histogram

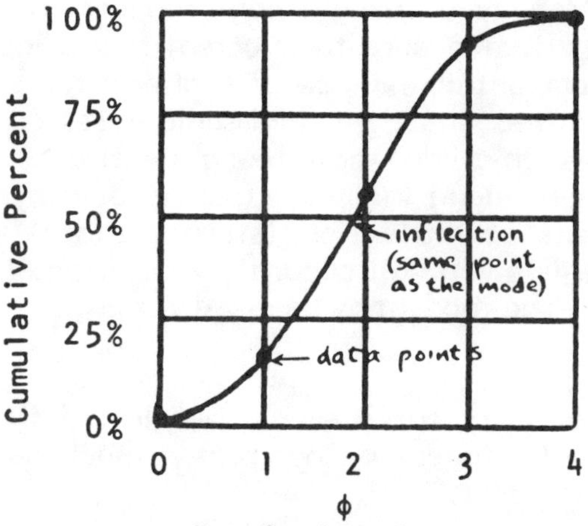

Cumulative Curve, Arithmetic Ordinate

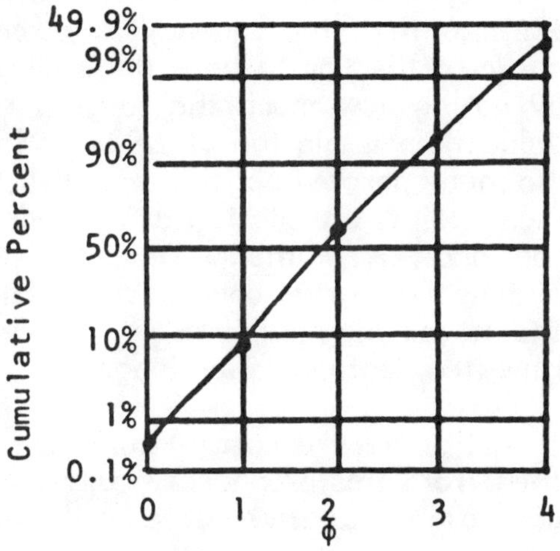

Cumulative Curve, Probability Ordinate

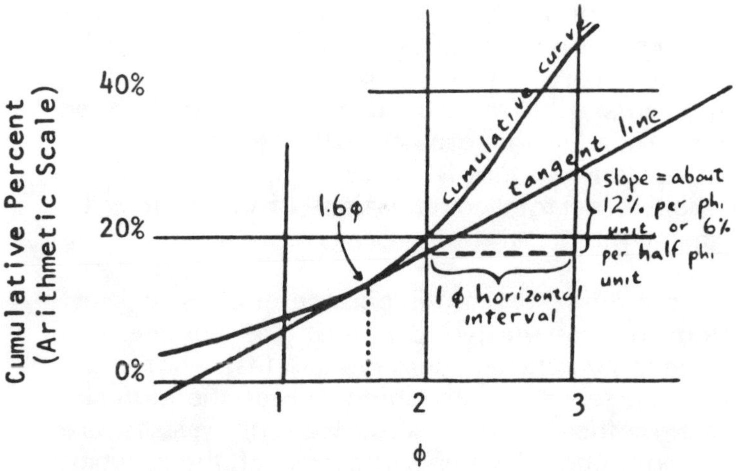

Method of construction. Here we are obtaining the value of the frequency curve at a diameter of 1.6φ. It is about 6% per half phi unit.

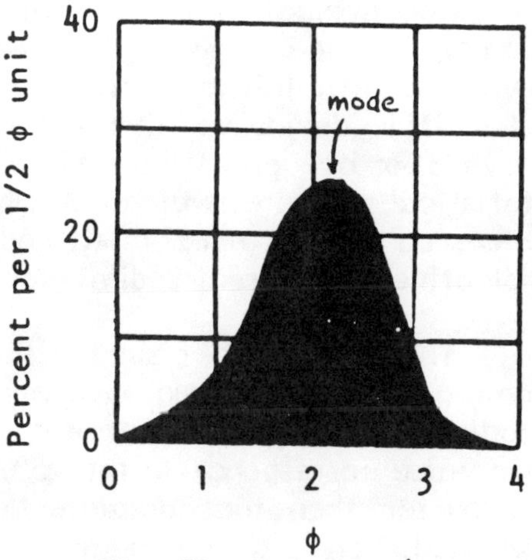

The final curve (not the same sample as the one at left).

Frequency Curve

extrapolation. HENCE, THIS IS THE CURVE THAT MUST BE USED FOR ALL DETERMINATION OF PARAMETERS. The only disadvantage is that it is even less pictorial than the arithmetic cumulative curve.

4. Frequency Curve. The frequency curve represents in essence a smoothed histogram in which a continuous bell-shaped curve takes the place of the discontinuous bar graph. Again, it is chiefly of pictorial value because no statistical parameters can be read for it. Although strictly pictorial, it gives a much better picture than this histogram because it is independent of the sieve interval used and is the best method to use in dissecting mixed populations into their separate normal distributions. Mathematically, it is the first derivative of the cumulative curve, and is thus obtained by measuring the slopes of the tangents to the cumulative curve. To construct it, one plots a cumulative curve with arithmetic (not probability) ordinates. Now one measures the slopes of tangents to this curve at various grain-size values. For example, if you want to find the frquency at a diameter of 2.83ϕ, lay a straight-edge tangent to the curve at the point where the 2.83ϕ line intersects it. Measure the slope of this tangent by noting how much the tangent rises over a horizontal distance of ½ phi unit. This value then is plotted at 2.83ϕ on the frequency curve, whose vertical scale is stated as "so many percent per half phi unit" (or half Wentworth grade--both being identical). As many points are plotted as needed, spaced anywhere along the curve (not necessarily at the analysis points). Be sure to get all points of inflection (steeper places on cumulative curve) and all "sags" (flatter places); thus you will accurately determine all the modes and minimums on the frequency curve. The mode may be fixed accurately by repeated approximation (page 42).

For very accurate work, a cumulative probability curve should be plotted first, then from this probability curve, data points may be taken to construct a much more accurate cumulative curve on the arithmetic graph paper.

Statistical Parameters of Grain Size

For evaluation of sets of samples it is probably best to compare the curves directly by eye as only in that way can the entire character of the sediment curves be revealed. But this is inconvenient and furthermore not very quantitative: it is often difficult to decide whether curve A represents a better sorted or finer sample than curve B, or how great the differences are. To solve this problem one resorts to various statistical measures which describe quantitatively certain features of the curves; these values can then be tabulated and, it is hoped, certain combinations of values may be indicative of different sedimentary environments.

There are two basic methods of obtaining statistical parameters. The most commonly used method involves plotting the cumulative curve of the sample and reading the diameter represented by various cumulative percentages (as, what grain size value corresponds to the 25% mark of the sediment--meaning 25% of the material is coarser than that diameter). In this method, much more accurate results are obtained if one plots the cumulative curve on probability paper, because of the superior accuracy of extrapolation and interpolation. The second method, called the method of moments, is far more complicated and probably of no greater value. It is explained at the end of this section.

The significance of most of the measures here described is shown graphically on page 51. Refer to it continually.

Measures of Average Size

It is desirable to have a measure which will say, "sample A is so much coarser than sample B." This is not nearly as easy as it looks, though, because there are many different measures of average size. Using one measure, sample A might be "coarser"; using another measure, sample B might be coarser. There is no consensus yet as to which is best. Until then, become familiar with all of them.

Mode (Mo) is the most frequently-occurring particle diameter. It is the diameter corresponding to the steepest point (point of inflection) on the cumulative curve (only if the curve has an arithmetic frequency scale). It corresponds to the highest point on the frequency curve. Several formulae have been developed for determination of the mode, but none of them are satisfactory. The only way the mode can be determined is by successive trials. Using the graph of the sample plotted on probability paper, one selects a point where the mode ought to be, and measures the percentage of the sample that occurs within the diameter range from $\frac{1}{4}\phi$ coarser than that point to $\frac{1}{4}\phi$ finer than that point (i.e., within $\frac{1}{2}\phi$ interval centered on the presumed modal point). Then he moves over a small distance (say 0.1 or 0.2ϕ) to a new presumed mode and measures the percentage occurring in the $\frac{1}{2}\phi$ interval centered on that new point. This is done repeatedly until the highest value is obtained which then corresponds to the modal diameter. It is often difficult to fix the mode more accurately than 0.1 or 0.2ϕ. Sediments not uncommonly have two or more modes, located by finding other points of inflection on the cumulative curve or other peaks on the frequency curve. Advantages: the mode is quite valuable in sediment genesis and transport studies, especially when two or more sources are contributing. The modal diameter often stays fairly constant in an area while the other, more "synthetic" measures tend to vary more erratically. It deserves more common use. The disadvantages are its lack of common usage, and in the fact that it is difficult to determine. Also, it is independent of the grain size of the rest of the sediment, therefore is not a good measure of overall average size.

Median (Md). Half of the particles by weight are coarser than the median, and half are finer. It is the diameter corresponding to the 50% mark on the cumulative curve and may be expressed either in ϕ or mm. (Md_ϕ or Md_{mm}). The advantage is that it is by far the most commonly used measure and the easiest to determine. The disadvantage is that it is not affected by the extremes of the curve, therefore does not reflect the overall size of sediments (especially skewed ones) well. For bimodal sediments it is almost worthless. Its use is not recommended.

Graphic Mean (M_z) (Folk). The best graphic measure for determining overall size is the Graphic Mean, given by the formula $M_z = (\phi16 + \phi50 + \phi84)/3$. It corresponds very closely to the mean as computed by the method of moments, yet is much easier to find. It is much superior to the median because it is based on three points and gives a better overall picture. This will be the standard measure of size used. Inman has used $(\phi16 + \phi84)/2$ as a measure of mean size but this is not satisfactory in skewed curves.

Measures of Uniformity

Several measures are available for measuring the uniformity or sorting of sediments. As a general rule, the more of the curve that enters into the sorting coefficient, the better the measure.

Trask's Sorting Coefficient (S_o) is used only with millimeter values, and is given by $\sqrt{Mm25/Mm75}$. Most beach sands have S_o = 1.3-1.5. In the past, it has been almost the only measure of sorting used, but its use is declining because it measures only the sorting in the central part of the curve. It should be abandoned.

Phi Quartile Deviation (QDϕ) is the exact analogue of S_o but adapted for the ϕ scale; it is given by (ϕ75-ϕ25)/2. This measure, like S_o, fails to give a good indication of sorting because they indicate only the sorting in the middle of the curve and ignore the ends, where the differences between samples are most marked. For example, a beach sand consisting of nothing else but fine and medium sand might have the same QDϕ and S_o as a sediment consisting of sand with 15% pebbles and 10% clay!

Therefore these measures should be no longer used.

Graphic Standard Deviation (σ_G) is (ϕ84-ϕ16)/2. It is very close to the standard deviation of the statistician (see the method of moments) but is obtained by reading two values on the cumulative curve instead of by lengthy computation. This sorting measure embraces the central 68% of the distribution, thus is better than QDϕ but not as good as σ_I. If a sediment has σ_G of 0.5ϕ, it means that two thirds (68%) of the grains fall within 1ϕ unit or 1 Wentworth grade centered on the mean - i.e., the mean $\pm$ one standard deviation.

Inclusive Graphic Standard Deviation (σ_I) (Folk). The Graphic Standard Deviation, σ_G, is a good measure of sorting and is computed as (ϕ84-ϕ16)/2. However, this takes in only the central two-thirds of the curve and a better measure is the Inclusive Graphic Standard Deviation, σ_I, given by the formula

$$\frac{\phi84-\phi16}{4} + \frac{\phi95-\phi5}{6.6}$$

This formula includes 90% of the distribution and is the best overall measure of sorting. It is simply the average of (1) the standard deviation computed from ϕ16 and ϕ84, and (2) the standard deviation as computed from ϕ5 and ϕ95-- since this interval (from 5 to 95%) embraces 3.30σ, the standard deviation is found as (ϕ95-ϕ5)/3.30. The two are simply averaged together (which explains why the denominators are both multiplied by 2).

Note that the standard deviation here is measure of the spread in phi units of the sample, therefore the symbol ϕ must always be attached to the value for σ_I. Measurement sorting values for a large number of sediments has suggested the following verbal classification scale for sorting:

σ_I under .35ϕ, very well sorted
.35-.50ϕ, well sorted
.50-.71ϕ, moderately well sorted
.71-1.0ϕ, moderately sorted
1.0-2.0ϕ, poorly sorted
2.0-4.0ϕ, very poorly sorted
over 4.0ϕ, extremely poorly sorted

The best sorting attained by natural sediments is about .20-25ϕ, and Texas dune and beach sands run about .25-.35ϕ. Texas river sediments so far measured range between .40-2.5ϕ, and pipetted flood plain or neritic silts and clays average about 2.0-3.5ϕ. The poorest sorted sediments, such as glacial tills, mudflows, etc., have σ_I values in the neighborhood of 5ϕ to 8 or even 10ϕ.

Measures of Skewness or Asymmetry

Curves may be similar in average size and in sorting but one may be symmetrical, the other asymmetrical. Skewness measures the degree of asymmetry as well as the "sign"--i.e., whether a curve has an asymmetrical tail on the left or right.

Phi Quartile Skewness ($Skq\phi$). This is found by (ϕ25 + ϕ75 - 2(Mdϕ))/2. A (+) value indicates that the sediment has an excess amount of fines (the frequency curve shows a tail on the right) and a (-) value indicates a tail in the coarse (left). The disadvantage of this measure is that it measures only the skewness in the central part of the curve, thus is very insensitive; also, it is greatly affected by sorting so is not a "pure" measure of skewness. In two curves with the same amount of asymmetry, one with poor sorting will have a much higher quartile skewness than a well-sorted sample.

Graphic Skewness. As a measure of skewness, the Graphic Skewness (Sk_G) given by the formula

$$\frac{\phi 16+\phi 84-2\phi 50}{(\phi 84-\phi 16)}$$

may be used (Inman). This measures the displacement of the median from the average of the ϕ16 and ϕ84 points (see figure below), expressed as a fraction of the standard deviation, thus the measure is geometrically independent of sorting. The derivation follows:

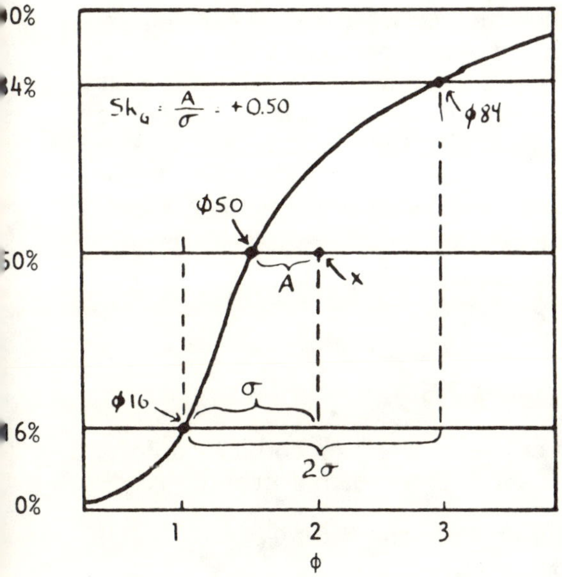

Let "x" be the midpoint of the ϕ16 and ϕ84 values, found by (ϕ16 + ϕ84) /2--in this case (1+3) /2 or 2.0ϕ. Then the distance "A" is the displacement of the Median (ϕ50) from the x_A midpoint. The skewness measure is then $\frac{A}{\sigma}$

$$\text{But } A = \frac{\phi 16+\phi 84}{2} - \phi 50,$$

$$\text{and} \qquad \sigma = \frac{\phi 84-\phi 16}{2},$$

so clearing fractions gives

$$\frac{\phi 16+\phi 84-2\phi 50}{(\phi 84-\phi 16)}$$

In this case, $\frac{1+3-2(1.5)}{(3-1)}$

or Sk_G = +0.50. Note that the median is displaced 0.50 of the way from the "x" midpoint to the ϕ16 or standard deviation mark.

Inclusive Graphic Skewness (Sk_I) (Folk). The skewness measure discussed above covers only the central 68% of the curve. Inasmuch as most skewness occurs in the "tails" of the curve, this is not a sensitive enough measure. A much better statistic, one that includes 90% of the curve, is the Inclusive Graphic Skewness given by the formula

$$\frac{\phi 16 + \phi 84 - 2\phi 50}{2(\phi 84 - \phi 16)} + \frac{\phi 5 + \phi 95 - 2\phi 50}{2(\phi 95 - \phi 5)}.$$

This formula simply averages the skewness obtained using the $\phi 16$ and $\phi 84$ points with the skewness obtained by using the $\phi 5$ and $\phi 95$ points, both determined by exactly the same principle. This is the best skewness measure to use because it determines the skewness of the "tails" of the curve, not just the central portion, and the "tails" are just where the most critical differences between samples lie. Furthermore it is geometrically independent of the sorting of the sample. Because in the skewness formula a measure of phi spread occurs both in numerator and denominator, the Sk_I value is a pure number and should not be written with ϕ attached. Skewness values should always be recorded with a + or - sign to avoid possible confusion.

Symmetrical curves have $Sk_I = .00$; those with excess fine material (a tail to the right) have positive skewness and those with excess coarse material (a tail to the left) have negative skewness. The more the skewness value departs from .00, the greater the degree of asymmetry. The following verbal limits on skewness are suggested: Sk_I from +1.00 to +.30, strongly fine-skewed; +.30 to +.10, fine-skewed; +.10 to -.10, near-symmetrical, -.10 to -.30, coarse-skewed; and -.30 to -1.00, strongly coarse-skewed. The absolute mathematical limits of the measure are +1.00 to -1.00, and few curves have Sk_I values beyond +.80 and -.80.

Measures of Kurtosis or Peakedness

In the normal probability curve, defined by the Gaussian formula, the phi diameter interval between the $\phi 5$ and $\phi 95$ points should be exactly 2.44 times the phi diameter interval between the $\phi 25$ and $\phi 75$ points. If the sample curve plots as a straight line on probability paper (i.e., if it follows the normal curve), this ratio will be obeyed and we say it has normal kurtosis (1.00). Departure from a straight line will alter this ratio, and kurtosis is the quantitative measure used to describe this departure from normality. It measures the ratio between the sorting in the "tails" of the curve and the sorting in the central portion. If the central portion is better sorted than the tails, the curve is said to be excessively peaked or leptokurtic; if the tails are better sorted than the central portion, the curve is deficiently or flat-peaked and platykurtic. Strongly platykurtic curves are often bimodal with subequal amounts of the two modes; these plot out as a two-peaked frequency curve, with the sag in the middle of the two peaks accounting for its platykurtic character. The kurtosis measure used here is the Graphic Kurtosis, K_G, (Folk) given by the formula

$$K_G = \frac{\phi 95 - \phi 5}{2.44(\phi 75 - \phi 25)}$$

This value answers the question, "for a given spread between the $\phi 25$ and $\phi 75$ points, how much is the $\phi 5$ to $\phi 95$ spread deficient (or in excess)?" For normal curves, K_G = 1.00; leptokurtic curves have K_G over 1.00 (for example a curve with $K_G = 2.00$ has exactly twice as large a spread in the tails as it should have for its $\phi 25$-$\phi 75$ spread, hence is much poorer sorted in the tails than in the central portion); and platykurtic curves have K_G under 1.00 (in a curve with $K_G = 0.70$, the tails have only 0.7 the spread they should have with a given $\phi 25$-$\phi 75$ spread). Kurtosis, like skewness, involves a ratio of spreads hence is a pure number and should not be written with ϕ attached.

The following verbal limits are suggested: K_G under 0.67, very platykurtic; 0.67-0.90, platykurtic; 0.90-1.11, mesokurtic; 1.11-1.50, leptokurtic; K_G over 1.50-3.00, very leptokurtic; K_G over 3.00, extremely leptokurtic. The absolute mathematical limits of the measure are from 0.41 to virtually infinity; few analyzed samples fall beyond the range from 0.60 to 5.0, however.

The distribution of K_G values in natural sediments is itself strongly skewed, since most sediments are around .85 to 1.4, yet some values as high as 3 or 4 are not uncommon. Thus for all graphic and statistical analysis (computation of mean or standard deviation of kurtosis, running of t tests, etc.) the kurtosis distribution must be normalized by using the transformation $K_G/(1 + K_G)$. Using transformed kurtosis (written K_G) a normal curve has a value of .50, and most sediments fall between .40-.65.

Characterization of Frequency Distribution

For characterization of the size frequency distributions of sediments, the limits suggested above should be followed. For size terms use the grain-size triangle (page 28). Here are some examples:

Fine sand, well-sorted, fine-skewed mesokurtic.

Sandy pebble gravel, moderately sorted, strongly fine skewed leptokurtic.

Granular medium sand, very poorly sorted, coarse-skewed very platykurtic.

Very fine sandy mud, very poorly sorted, near-symmetrical platykurtic.

These various statistical measures may be plotted against each other to see how, for example, skewness values may be related to mean size (although geometrically independent, in any given set of samples the two values may show some correlation). They may be plotted on recent sediment maps and contoured to show the regional variation of the measures, and provide a clue to identification of ancient environments.

Frequency distributions of other sets of data may be statistically analyzed in exactly the same way as grain size. The measures of mean size, standard deviation, skewness and kurtosis used here may be used for any type of data at all, in any field of science. The verbal limits for skewness and kurtosis suggested here may also be used for data in any other field, but the verbal limits on size and standard deviation are of course inapplicable.

The Method of Moments

The second method of obtaining statistical parameters is called the method of moments. It is a computational (not graphical) method of obtaining values, in which every grain in the sediment affects the measure. Thus it probably gives a truer picture than the graphic methods, which rely on only a few selected percentage lines. For example, the median is obtained graphically by merely reading the diameter at the 50% mark of the cumulative curve, and is not at all affected by the character of the rest of the curve; but the mean, computed by the method of moments, is affected by the distribution over every part of the curve (see sheet at the end giving graphic significance of measures). Details on the computations involved are given in Krumbein and Pettijohn, and many computer programs are available for calculation of these values very rapidly. It is possible to obtain skewness and kurtosis also by means of moments, but we will confine ourselves to determination of the mean and standard deviation. To begin with, one sets up the following form, using the ϕ scale. A midpoint of each ϕ class is selected (usually 2.5, 3.5ϕ) but in some cases (as in Pan fraction) a different midpoint must be selected. In this "open-ended" distribution, where there is a lot of material in the pan fraction of unknown grain size, the method of moments is severely handicapped and probably its use is not justified.

ϕ class interval	ϕ mid-point	Weight, grams	Product	Midpoint Deviation	Mid. Dev. squared	Product
	(D)	(W)	(D.W.)	($M\phi$-D)	$(M\phi\text{-}D)^2$	$W(M\phi\text{-}D)^2$
0.0 - 1.0	0.5	5.0	2.5	2.18	4.74	23.7
1.0 - 2.0	1.5	10.0	15.0	1.18	1.39	13.9
2.0 - 3.0	2.5	30.0	75.0	0.18	0.03	0.9
3.0 - 4.0	3.5	20.0	70.0	0.82	0.67	13.4
4.0 - pan	5*	5.0	25.0	2.32	5.39	27.0
sum (Σ)	--	70.0	187.5			78.9

* arbitrary assumption.

The phi arithmetic mean of the sample ($M\phi$) is then

$$\frac{\sum DW}{\sum W} = \frac{187.5}{70.0} = 2.68\phi.$$

The "midpoint deviation" column is then obtained by subtracting this mean (2.68) from each of the phi midpoints of each of the classes. These deviations are then squared, multiplied by the weights, and the grand total obtained. Then the standard deviation ($\sigma\,\phi$) is obtained by the following formula:

$$\sigma_\phi = \sqrt{\frac{\Sigma\left[W(M\phi - D)^2\right]}{\Sigma\, W}} = \sqrt{\frac{78.9}{70}} = \sqrt{1.127} = 1.06\,\phi$$

Special Measures

This whole problem may be approached fruitfully in another way. Instead of asking, "what diameter corresponds to the 50% mark of a sample" we can ask "what percent of the sample is coarser than a given diameter." This is an especially valuable method of analysis if one is plotting contour maps of sediment distribution in terms of percent mud, percent gravel, percent of material between 3ϕ and 4ϕ, etc. It works better than any other method if the sediment is bimodal. Or, one may plot as a contour map the diameter of the largest particle in the sample. Try new methods all the time.

Measures of Average Size and Standard Deviation

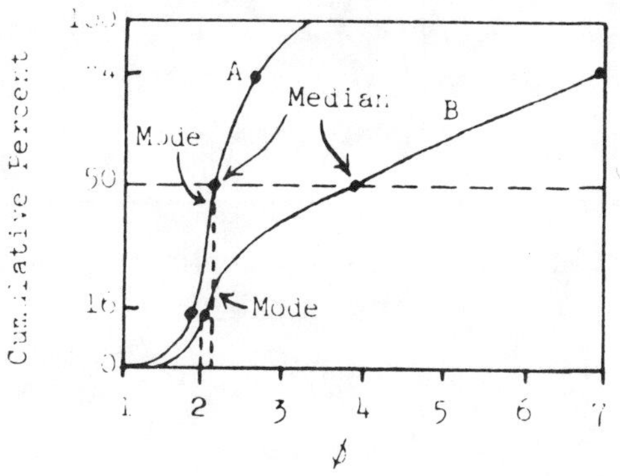

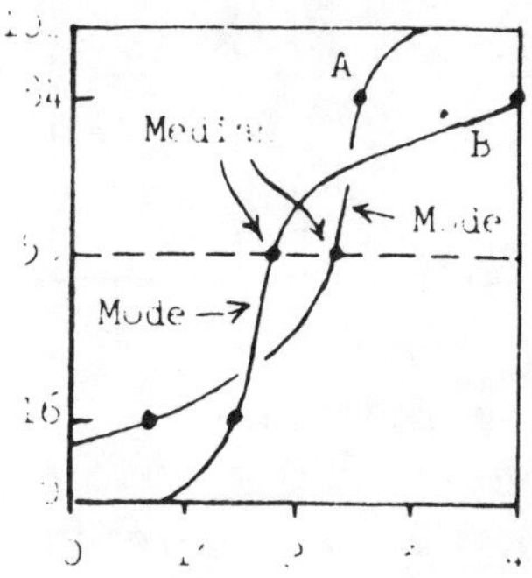

Mode, Median and Mean. In the left-hand figure, both curves have a mode at about 2ϕ, but B has a much finer median. The Graphic Mean (which is the average of the three heavy dots at the 16, 50, and 84 percentiles) is about 2.2ϕ for A and about 4.2ϕ for B. In the right-hand figure, B has a much coarser median and mode, but actually is finer if you consider the entire sediment; the Graphic Mean for A would be about (0.7 + 2.4 + 2.6)/3, or 1.9ϕ, and for curve B would be about (1.4 + 1.7 + 4.1)/3, or 2.4ϕ. The Mean gives the best overall indication of average size; the mode is of real value, but the median is not a good measure.

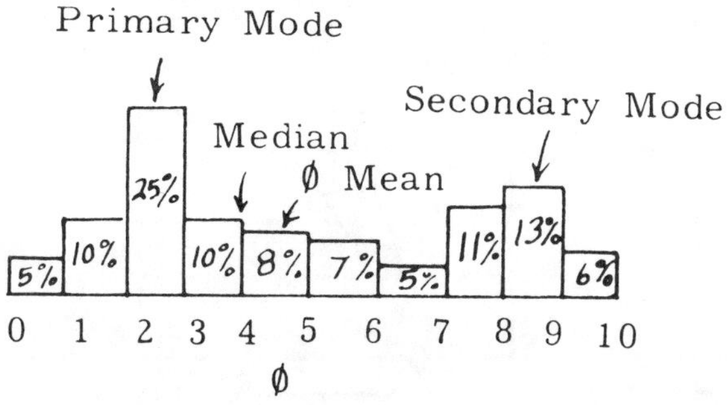

This histogram illustrates the difference between the several measures of average size. Note there are two modes, at about 2.5ϕ and 8.3ϕ. The median has half the sample finer and half the sample coarser than itself; the left four columns total 50% of the sample so the median is 4.0ϕ. The ϕ Mean, computed by the method of moments, is based on the entire curve (5 x 0.5ϕ + 10 x 1.5ϕ + 25 x 2.5ϕ. . . . . 13 x 8.5ϕ + 6 x 9.5ϕ)/100; the ϕ Mean is thus 4.7ϕ. The Graphic Mean, which is a rapid and close approximation to the ϕ Mean, is also 4.7ϕ but this must be found on the cumulative curve. Changing the 6-7ϕ grade to 16% and the 8-9ϕ grade to 2%, will affect the means but will not at all affect the primary mode or median (prove this for yourself).

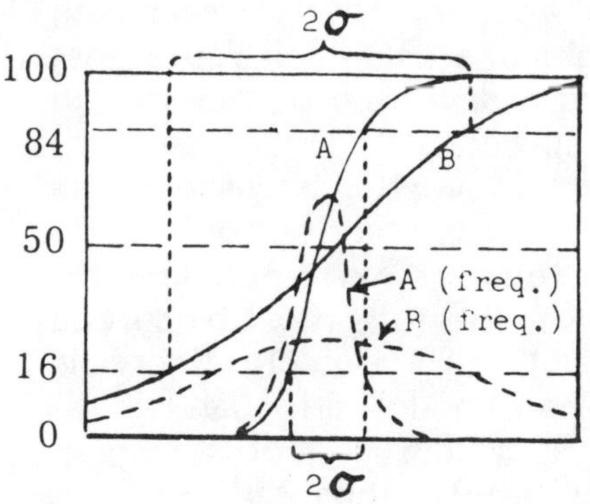

Cumulative and Frequency (dashed) curves for well-sorted sediment (A) and poorly sorted sediment (B). In the well-sorted sediment, the cumulative curve is much steeper, the frequency curve is much higher and much less spread out, and the range between 16 and 84 percentiles (which equals 2σ) is much smaller, hence the standard deviation is much smaller.

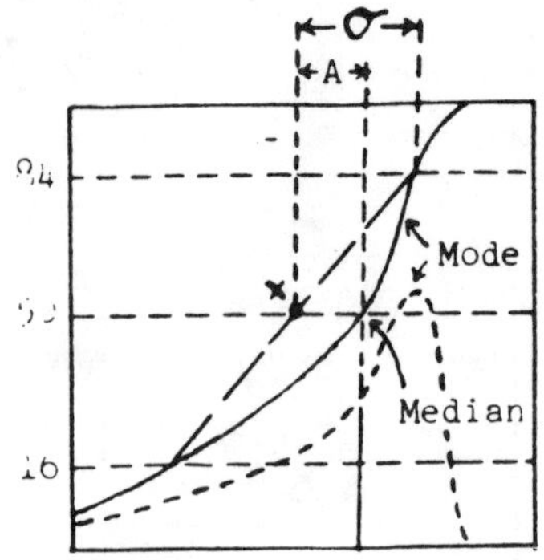

Skewed with excess coarse material (tail in the coarse, or left side). X is the midpoint of the 16 and 84 marks;

$$Sk_G = \frac{A}{\sigma} = -.60$$

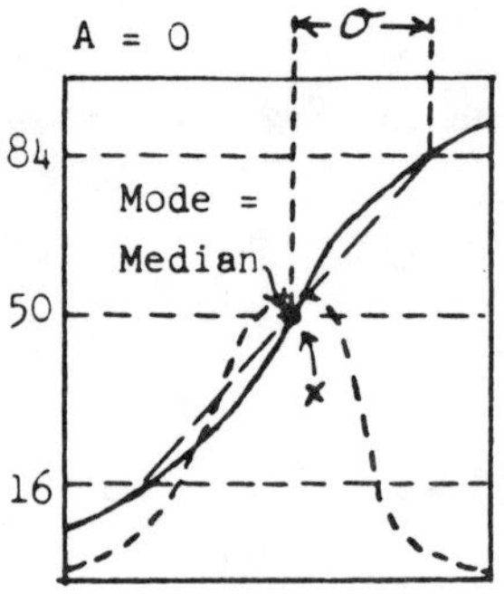

Symmetrical. Mode equals median.

$$Sk = .00$$

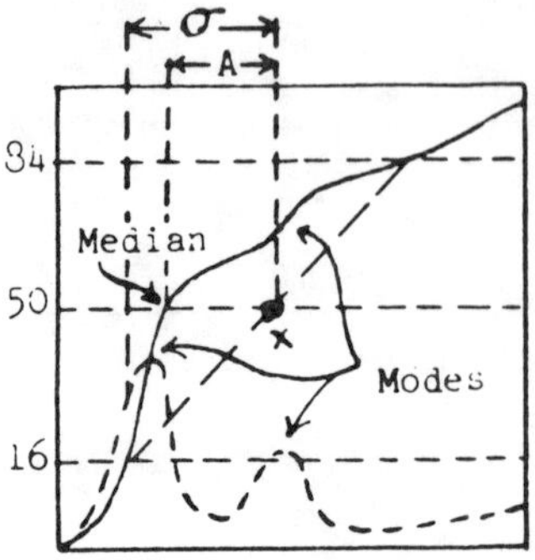

Skewed with excess fine material (tail in the fine, or right side); also bimodal.

$$Sk_G = \frac{A}{\sigma} = +.75$$

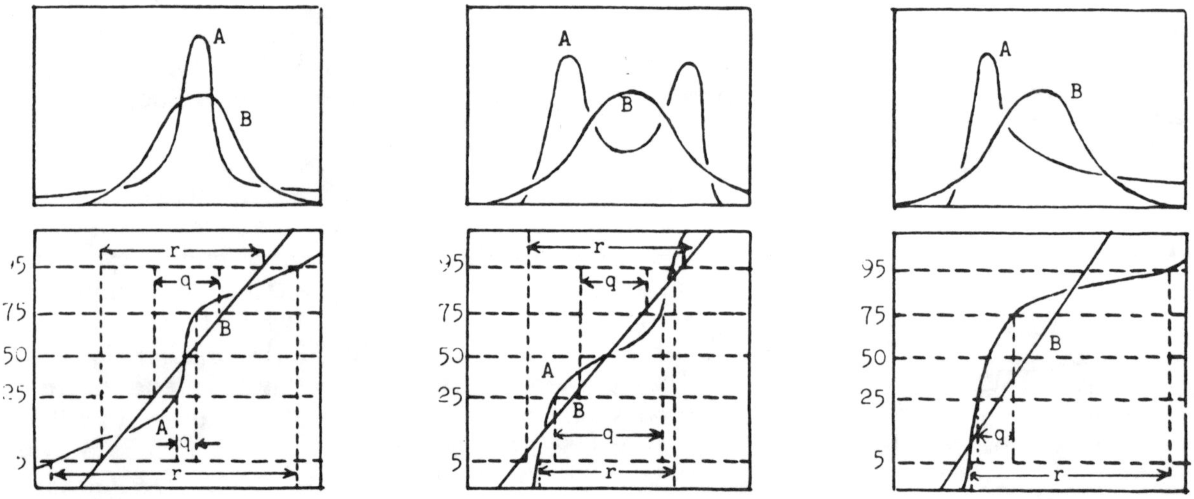

Kurtosis. All the cumulative curves are drawn on probability paper. The corresponding frequency curves are shown above. Each graph has a straight line (B) which represents a normal probability curve with the same mean and standard deviation as the sample curve (A). In a normal curve, the ratio r/q (r is the 5-95 spread, q the 25-75 spread) must be exactly 2.44. The left-hand figure is extremely leptokurtic, because r/q is much greater than 2.44. Note the frequency curve which is excessively peaked at the middle, is deficient in the "shoulders" of the curve, and the tails are extremely long (to get the same overall spread as the corresponding normal curve, the tails must be spread wider to compensate for the condensation of the center.). The middle figure is extremely platykurtic, with r/q much less than 2.44; it is also bimodal, although not all platykurtic curves are bimodal. The right-hand figure is a curve that is highly leptokurtic and also has strong positive skewness; r/q is much greater than 2.44.

A FEW STATISTICAL MEASURES FOR USE IN SEDIMENTARY PETROLOGY

Introduction. A good knowledge of statistics is becoming essential for anyone who wishes to work in any of the sciences, because the whole of scientific work from laying out the experiment to interpretation of data is based on statistics. Trying to use numerical data without a knowledge of statistics is like trying to drive without a brake. You never know where you will end up and the odds are you will end up in the wrong place and get the wrong conclusion.

In sedimentary petrography, statistics are used in laying out the sampling program; in determining the best experimental technique for analysis; in collecting the analytical data; and in drawing correct geological conclusions, such as: what is the content of feldspar in X formation? Within what limits am I certain this value is correct? What is the spread of values to be expected? Does X formation have more feldspar than Y formation, and how confident am I of this? Does its heavy mineral content differ significantly from that of formation Y? What is the relation between grain size and zircon content, expressed mathematically?

This outline is not intended to make you an expert in statistics. It merely shows examples of the use of statistics in petrography, with the hope that it will stimulate you to take several courses or read up on your own. It is super-simplified and condensed, therefore omits a lot of material that should really be covered. For further information refer to any standard textbook, especially for geologists: Miller and Kahn, 1962; Krumbein and Graybill, 1965: Griffiths, 1967; Koch & Link, 1972; Davis, 1973.

The Normal Probability Curve. In order to understand some of the assumption and underlying principles, it is essential to study the statisticians' most fundamental concept, that of the normal probability curve. This is the basis for study of experimental data of all kinds.

As a first step in the analysis of data from any field of science, one usually constructs a frequency distribution. For example, if one is studying the batting averages of baseball players, he would select convenient class intervals to divide the entire range of data into about 10 to 20 classes and proceed to find how many batters had averages between .200 and .210, how many between .210 and .220, and so on; here the class interval would be .010. Or if an anthropologist were studying the lengths of human thigh bones, he would first ascertain the spread between the largest and smallest bone (say for example 12" to 31"), and divide this into a convenient number of classes. Here a convenient class interval would be 1", and he would proceed to find how many thighbones were between 12" and 13", how many between 13" and 14", and so on. When data of this type is plotted up in the form of a histogram or frequency curve, it is usually found that most of the items are clustered around the central part of the distribution with a rapid "tailing off" in the extremes. For example, far more baseball players hit between .260 and .280 than hit between .320 and .340, or between .180 and .200. Even less hit between .100 and .120, or between .380 and .400. A great many types of data follow this distribution, and the type of frequency curve resulting is called the "probability curve," or the "normal curve," or often a "Gaussian curve" after Gauss who was a pioneer in the field. The curve is defined as the kind of distribution resulting if one had 100 well-balanced coins and tossed them all repeatedly to count the number of heads appearing. Naturally, the most frequent occurrence would be 50 heads and 50 tails; 40 heads or 60 heads would be less likely, and so on down to occurrences of 10 heads and 90 tails, which would be very few; a throw of 1 head and 99 tails would be exceedingly rare. By tedious computations, one could figure the chances of throwing any combination of heads and tails, and this is the useful feature of the normal probability curve: the probabilities fall off at a definite, predictable rate which is fixed by a mathematical equation. Furthermore, the curve is symmetrical about the mean--a throw of 34 heads (50-16) is exactly as likely as a throw of 66 heads (50 + 16); and a throw of 18 heads is just as likely as 82 heads.

Many types of data follow closely this curve which is defined by coin-tossing experiments and rigidly fixed by an equation. Baseball batting averages; weights of bolts turned out by a factory; life spans of electric light bulbs; mean daily temperatures for any month if records are kept over some years; heights of people; densities of granite samples; widths of brachiopod valves; slope angles of geomorphic features and many others often follow this normal probability curve, providing enough data is collected. For example if one chose ten people and weighed them his curve would be rather irregular; by the time he weighed 1000 people it would be much smoother, and if he weighed 10,000 the distribution would hew very closely to the normal curve.

Many distributions do not follow the theoretical normal distribution, however. One of the most common ways that a distribution departs from normality is in its lack of symmetry. The graph of a normal distribution is a perfectly symmetrical bell-shaped curve, with equal frequencies on both sides of the most common value (i.e. in the 100-coin toss, 25 heads are just as common as 75 heads). Many kinds of data are asymmetrical, though. Consider the prices of houses in an average American city. The most common price might be somewhere around $10,000. In this city there might be many $25,000 homes; in order to have a symmetrical frequency distribution, this would demand that there be many homes that cost minus $5000! This curve, then would be highly asymmetrical with the lowest value being perhaps $3000, the peak frequency at about $10,000, and a long "tail" in the high values going out to perhaps $100,000 or even more. The distribution of the length of time of long-distance telephone calls is also a distribution of this type, since most calls last between two and three minutes, very few are less than one minute but there are some long distance calls lasting as long as 15 minutes or even an hour. The frequency distribution of percentage of insoluble materials in limestone samples is also a highly asymmetrical or skewed distribution, with most limestones in one formation having, for example, between 5 and 10 percent insoluble, but some samples having as much as 50 or 75% insoluble, with 0% as the obvious minimum percentage.

A further way in which distributions depart from normality is that they may have two or more peak frequencies (termed modes). If one took a large college building and obtained the frequency distribution of ages of all people in the building at a given time, he would find a curve that had two peaks (bimodal) instead of one. The highest peak would be between 19 and 20 (the average age of students who would make up most of the population), and another peak might occur at around 40 (the average age of the professors), with a minimum at perhaps 30 (too young for most professors, and too old for most students). This distribution would be distinctly non-normal; technically, it would be said to have deficient kurtosis. In geology we would obtain a similarly non-normal, bimodal distribution if we measured the sizes of crystals in a porphyritic granite, or if we measured the percentage of quartz in a sand-shale sequence (sand beds might be almost pure quartz, while the shale beds might have 20% or less).

In analyzing frequency distributions, the most common measures used are the arithmetic mean (or average) and the standard deviation (or degree of scatter about the mean). Skewness and kurtosis are also used for special purposes (see pages 45-46).

Arithmetic Mean: The most common and useful type of average for ordinary purposes is the arithmetic mean, symbolized $\bar{x}$. To obtain this value, simply add up the values and divide the result by the number (n) of values. This is symbolized as

$$\bar{x} = \frac{\sum x}{n} .$$

There are a number of other averages (median, mode, geometric mean, etc.) which are useful in special cases--especially when the data is skewed, as when you are taking an average consisting of a lot of small values and a few very large values, which would tend to distort the simple arithmetic mean.

Standard Deviation: For most data we are interested in determining the spread or scatter of the values. For example, a set of reading 4, 6, 7, 4, 5, 6, obviously is less scattered than a set such as 2, 5, 14, 1, 9, 3, although both have the same mean. The standard deviation (s or σ) is a precise measure of this scatter and, next to the mean, is the most useful value in statistics. It is the only measure of spread that has wide use. The standard deviation is computed such that, on an average, 68% of the samples will fall within plus or minus one standard deviation ($\pm \sigma$) from the mean; 95% will fall within plus or minus two standard deviations ($\pm 2\sigma$); and 99% of the samples will fall within $\pm 3\sigma$. For example, if we have 100 porosity values on a sandstone formation and the mean porosity is 15.0% with a standard deviation of 3.0%, we know that approximately 68 of these 100 samples will have porosities ranging between 12 and 18% ($\bar{x} \pm \sigma$), that 95 samples will range between 9 and 21% ($\bar{x} \pm 2\sigma$), and that 99 samples will range between 6 and 24% ($\bar{x} \pm 3\sigma$). Of course this assumes that we have a sufficient number of values, and that the values follow the normal bell-shaped probability curve (in which most of the values cluster about the mean, and "tail off" symmetrically to each side). This is tacitly assumed for most statistical work. A complete table of the percent of values included in a number of standard deviations is given on page 61.

To obtain the standard deviation there are two different methods which give identical answers. In the long method, one finds the arithmetic mean of his values, then finds the deviation of each value from this mean, and squares these deviations; then adds up these squared values, divides by the number of values and takes the square root of this quotient. In the short method, which we will use, one substitutes values in the following formula:

$$s = \sqrt{\frac{\sum(x^2) - \frac{(\sum x)^2}{n}}{n-1}}$$

For example, let us say that we have five thin sections of a formation and wanted to know the feldspar content. By point counter we obtain the following percentages on each slide: 8, 11, 6, 15, 10. What is the mean and standard deviation? We set up the following table:

	x	x^2
	8	64
	11	121
	6	36
	15	225
	10	100
Σ =	50	546

n, the number of samples, is 5; Σx (the sum of the x values--in this case the percentages of feldspar) equals 50, therefore the arithmetic mean, $\bar{x}$, is:

$$\frac{\sum x}{n} = \frac{50}{5} = 10.0\%$$

To find the standard deviation, we substitute values in the equation above.

(x^2), the sum of the squared values of x, equals 546; $(\Sigma x)^2/n$ is $50^2/5$ or 500; and n-1 is 5-1 or 4. Substituting,

$$s = \sqrt{\frac{546-500}{4}} = \sqrt{\frac{46}{4}} = \sqrt{11.5} = 3.4\%$$

If our samples are representative and the feldspar content follows the normal bell-shaped probability curve, then, if we collected 100 samples of the same formation, we would expect 68 of them to have feldspar contents between 6.6 and 13.4%. If someone asks you, "what are the odds of finding a sample with over 17% feldspar, you can immediately answer "only 2 or 3 out of 100." Why? Well, 17% is about equal to x + 2s, or 10.0 + 2(3.4); we know that 95% of the samples will fall in the range of $\bar{x} \pm 2s$, therefore only 5% of the samples will fall outside this range; of these, half will be higher and half lower, so 2.5% of the samples will have less than 3% feldspar, and 2.5% will have over 17%. Again, this assumes a normal distribution symmetrical about the mean.

Confidence Limits on a single mean. In the example above, we got a mean of 10.00 percent feldspar based on five samples of the formation. This is our estimate, based on five samples, of the feldspar content of the entire formation. Now the true feldspar content could not be determined unless we analyzed every one of the millions of sand grains in the entire formation. This we cannot do, so we have to estimate the mean by taking a small number of representative samples. But by the test to be described, we can tell how close our estimated mean is likely to be to the true mean of the formation--in other words, we can assign confidence limits to the mean. We can say, "I am 95% sure that the true formation mean lies between 9.0 and 11.0%,"--which means we will be wrong only 5% of the time--or one time in twenty.

We realize immediately that if we take 100 samples we will be more confident of our mean, than if we took only five samples. Also, we will be more confident of the mean if the values show a small spread (standard deviation) than if they show a large spread (Example: if we have values of 4.2, 3.6, 4.0, 4.8, 3.4 we are more confident that our mean is nearly 4.0 than if we have values 1.2, 4.8, 7.6, 3.1, 3.3). The formula for computing the confidence limits appropriately then takes these two factors into account.

$$\pm L = \frac{ts}{\sqrt{n}}$$

where n is the number of observations or values, s is the standard deviation, and t is explained below.

The factor t is put into the equation so that we can choose our "confidence level." Assume a formation with a mean porosity of 12.0%, standard deviation 3.0%. If we want to be only 50% sure of our mean, we look up in a table the value of t at the 50% level, and insert this in the formula. Let us say this makes "L" come out to ± 1.5. This means that we would be only 50% sure that our true formation mean lay between 10.5 and 13.5%, and there is one chance in two that the true mean lies either above or below these limits. However, if we chose the 95% level, we insert this value of t in the formula. Say that L then comes out to 3.0%; then we would be 95% sure that the true formation mean lay between 9.0 and 15.0%, and there would be only one chance in twenty of our being wrong. Most statisticians use the 95% level all the time. (P = 0.05 column, page 62).

The value t is found in a table with the levels to be selected (5%, 10%, 20%, etc.) plotted against the "degrees of freedom" (which in this case simply means the number of values minus 1); the value of t to be used lies at the intersection of our chosen confidence level and the degrees of freedom. For example, with 21 values, at the 5% (.05) level, t = 2.09, as shown in t table on page 59.

Populations and Probability

In the tests that follow it is essential that we understand the two concepts of "populations" and "probability."

Population is the term used for the data we are sampling; if we are measuring porosities, the "population" that we are sampling is the vast number of porosities present in every cubic inch of that formation, of which we test what we assume are a representative few; if we are measuring the mica content of a bed, the "population" is the percent of mica in each minute part of that bed; if we fill a jar with 500 black and 1,000 red beans and proceed blindfolded to pull out 50 beans, then we hope that our sample of 50 is representative of the true "population"--i.e., the 1,500 total beans. The population is the vast amount of numerical data available, of which we take only a small sample.

In many statistical tests, we try to answer the question, "what are the odds that we could have obtained as great a difference or greater by chance sampling of the same population?" This gives us an insight whether or not there is a real difference in the properties of the two formations we have sampled, or whether we could have gotten just as large a difference in, say chert content, by chance sampling of a single formation (a "homogenous" population). Let's think about it this way. We have a square jar and a round jar, each filled with a certain mixture of 1,000 black and white beans. The round jar has a population with a certain number of blacks, the square jar has a different proportion of blacks. You are now blindfolded and asked to pull 50 beans from one of the jars; let's say you got 30 blacks and 20 whites. Now, still blindfolded, you are asked to reach again into a jar (you still don't know which one you are reaching into) and pick 50 more beans. This time you get 22 blacks and 28 whites. Statistics enables us to anwer the question, "on the second drawing, did I take from the same jar as I did the first time; what is the probability that I drew both sets of beans from the same jar?" Or phrased differently, "what are the odds that I drew both samples of 50 from the same population (i.e., the same jar?)" This is essentially what we do when we obtain numerical data from rocks. We take only a small sample of the numerical data available in the formation, then sample another formation and see if there is any difference between the two formations--and difference in the population of beans in the jar, so to speak.

Probability. Many statistical tables have a critical value called "P" as an important part. P stands for probability--in most cases, P, represented as a percent, stands for the probability of a certain event happening. In the jar experiment above, after running through the computations of the statistical test, we enter a table and come out with a certain value for P. In this case let us say P came out to be .10 (10%). This means that if we had repeatedly sampled the same jar, only once in every ten times would we have gotten as large, or larger, a difference as we did on the two draws. Or, there is only one chance in ten that we have sampled the same population. Although it is not technically correct to say it this way, we may state the corollary that there is a 90% chance that we have sampled two different jars. Most statisticians do not accept an experiment as significant unless P reaches the 5% level (.05 on the table, pages 59 and 60).

Comparison of two means. We have studied the heavy mineral suites of two formations. We examine 50 samples of the Abner formation and find that within the heavy minerals, of the garnet content averages 15.0% with a standard deviation of 5%; and 30 samples of the Benjamin formation show a garnet content of 10.0 with a standard deviation of 4.0%. The question arises, does the Abner really have more garnet than the Benjamin, or could such differences have arisen by chance sampling of a homogeneous population? The so-called "t test" has been devised to answer this question and is one of the most useful statistical devices. We substitute our values in the following equation, where $\bar{x}$ is our mean garnet content, and n equals the number of samples (subscript $_a$ refers to values from the Abner formation, $_b$ to the Benjamin formation).

$$t = \frac{\bar{x}_a - \bar{x}_b}{s} \sqrt{\frac{n_a \cdot n_b}{n_a + n_b}}$$

where s approximately equals the average standard deviation of the two sets of values, and is found by the following equation (which is merely an expansion of our regular formula for the standard deviation):

$$s = \sqrt{\frac{\Sigma(x_a^2) - \frac{(\Sigma x_a)^2}{n_a} + \Sigma(x_b^2) - \frac{(\Sigma x_b)^2}{n_b}}{n_a + n_b - 2}}$$

We come out with a certain value for t. From here we enter a table (p. 59) showing t values as a function of the number of "degrees of freedom" and of P (probability). To find the degree of freedom in this case, we add up the number of values in both sets of data and subtract 2 from the result (in the example, 50 + 30-2 = 78 d.f.) Entering the table with 78 degrees of freedom, we read across the horizontal row at this value until we encounter the correct t value. From this t value we read straight up to see what P corresponds to our t. If P lies between .05 and .10, then we know that there is something between a 5% and 10% chance of our obtaining such differences (or larger) by chance sampling of a homogeneous population. This may be colloquially stated as saying "our experiment has shown that there are only 5 to 10 chances in 100 that the garnet content of the two formations is the same," or the reverse, "there is a 90 to 95% chance that the Abner really does contain more garnet than the Benjamin." Both these statements are technically not exactly precise, but may be considered as pretty close to the truth.

As stated before, statisticians ordinarily consider that in any experiment that fails to reach the 5% level, the data do not warrant making a conclusion. In other words if your P comes out to the 10% level, it means that you have failed to find a really significant difference between the two formations, either because the difference in means is too small, or the standard deviation is too large, or you took too few samples. The only way to remedy this situation is to take enough samples to push the results beyond the 5% level.

The X^2 (Chi square) test. The t test is used when you are comparing means of measurements (like grain size, percentages, porosities, densities, etc.) between two formations and can generally be done on only one property at a time.

The X^2 test is used when you are comparing counts of discrete objects between formations (like the number of grains larger than a given size in formation; or the number of grains of a certain heavy mineral; or in ordinary life, the number of accidents in a given city each month; or the number of home runs hit by a team each week). In some cases it is possible to use both the X^2 test or the t test, as in heavy mineral counts (which may be converted to percentages); but the X^2 test has this advantage, that more than one item or property can be compared at a time. For instance, if we count the number of tourmaline, rutile and zircon grains in two formations, we can use this whole stack of data at once with X^2 test and see if the formations are significantly different. If we used the t test we'd have to make a separate test for the tourmaline, another one for the zircon, another one for the rutile and there would be no simple way to combine them.

To use the X^2 test we must in some way arrive at a predicted or expected frequency or occurrence, based on a prior reasoning or else on long experience. We then compare our observed frequency with the given frequency. For example, if we roll a dice 600 times, we would expect that if the dice is well balanced each number (1 through 6) would appear equally and our expected frequency would therefore be 100. When we actually do roll the dice, we may find that we got only 95 1's, 123 2's, 82 3's, etc. The purpose of the X^2 test is to compare the observed with the expected frequency and see how likely it is that the deviations are due to chance, or whether the dice really is loaded. As another example, say that data collected over a period of 20 years shows that American League teams hit a grand total of 1,035 homers per season, on the average; thus we would expect them to hit about 1,035 homers this season. Well, this season they actually slammed 1,230 round trippers; is this "deviation" just due to chance, or to a new "lively" ball? The X^2 test helps us to answer this question.

There are a number of different ways to set up the X^2 test. The one we'll use most is called a "2 by n" table--usually we are comparing two formations based on counts of "n" types of heavy minerals, where "n" may be anything from 1 to 20 or more. For example, let us say that we count varieties of tourmaline in the Eli and Nineteenten formations and want to know if a significant difference exists between them in this respect. We tabulate the number of grains thus:

	Eli fm.	Nineteenten fm.	Total
Green tourmalines	15	25	40
Brown tourmalines	10	12	22
Other colors	20	14	34
Total	45	51	96

(This is a 2 by n table where n, the number of different mineral types equals 3)

Does this data indicate that there is a real difference in the types of tourmaline between the two formations? Or could we have gotten counts as different as this by chance sampling of formations that were identical in tourmaline varieties?

Please remember, to use the X^2 test, data must be stated in terms of actual numbers of grains counted--you cannot convert to percentages and then use the X^2 test.

We start out with the assumption that both have the same heavy mineral content and proceed to compute our expected frequencies on this basis. To figure the number of green tourmalines expected in a count of 45 grains of the Eli formation, we decide (from looking at the totals) that 40/96 of the grains should be green; thus we expect 45 times 40/96 green grains in the Eli, or 18.7. Since there were a total of 40 green grains counted, then we'd expect 40-18.7 or 21.3 green grains to occur in the Nineteenten.

The expected number of browns is 45 times 22/96 or 10.3 for Eli, and 22-10.3 or 11.7 for Nineteenten. For "others" we expect 45 times 34/96 or 15.9 for Eli, and 34-15.9 or 18.1 for Nineteenten. Now we set up the following table, subtracting 0.5 from each of the differences, giving our corrected differences (the reason we subtract 0.5 from each difference is a "correction for continuity" because obviously you cannot count fractional grains, and 10 grains is as close as you can possibly get to an expected frequency of 10.3 grains).

		Eli	Nineteenten
	Observed	15	25
Green	Expected	18.7	21.3
	Difference	3.7	3.7
	Corr. Diff.	3.2	3.2
	Observed	10	12
Brown	Expected	10.3	11.7
	Difference	0.3	0.3
	Corr. Diff.	0.3	0
	Observed	20	14
Other	Expected	15.9	18.1
	Difference	4.1	4.1
	Corr. Diff.	3.6	3.6

For each of the six "cells" (three tourmaline varieties in each of two formations) we square the corrected difference and divide it by the expected frequency <u>for that one cell</u>. Then we <u>add up these values</u> for the entire six cells and the total gives us the value of X^2, in this example 2.56. Symbolically, this operation is

$$X^2 = \Sigma\left(\frac{D^2}{E}\right)$$

$$X^2 = \frac{10.24}{18.7} + \frac{10.24}{21.3} + \frac{0}{10.3} + \frac{0}{11.7} + \frac{12.95}{5.9} + \frac{12.95}{18.1} = 2.56$$

We enter a X^2 table (p. 60) and must know two things: our value for X^2 and the degrees of freedom (d.f.). This time the d.f. is obtained by multiplying these two quantities: (number of horizontal data columns minus one) times (number of vertical data columns minus one); in this case (3-1)(2-1), or two degrees of freedom. Again, we read across the horizontal row corresponding with the proper number of degrees of freedom until we find our value of X^2 in the body of the table. Then we read directly up to the top of the table to find the corresponding "P". This will answer the question, what are the chances (out of 100) that such differences--or larger--would be obtained in random sampling of two uniform formations (or in sampling a homogeneous population?) Our result came out P = approximately .20, in other words there is one chance in five that we have sampled a homogeneous population, or conversely, four chances in five that the formations have a differing tourmaline content--though this last statement is not strictly true.

For another example, consider a tire company that for years has averaged 31 blowouts per million tire miles. After switching to a new type rubber, they find 44 blowouts occur in the next million miles. Is the new rubber inferior, or is this merely an expectable chance fluctuation? Using the X^2 test, the expected frequency would have been 31 blowouts for the new tires. (0-E) is 44-31 or 13, and correcting for discontinuity gives 12.5. $X^2 = (12.5)2/31 = 156/31 = 5.0$. In the table, for $X^2 = 5.0$ and d.f. = 1, P = .03; hence it is 97% certain that the new rubber is inferior.

There is one serious caution about using the X^2 test. In case the expected frequency in any cell is less than 5, this cell must be combined with another to bring the total expected frequency for the combined cells over 5. In the sample above, let's say I also counted 4 orange tourmalines and 3 yellow ones; in order that no expected cell frequency be under 5, I would have to lump these with other rare types in the cell labeled "others."

Other Techniques. Any statistical text will list many other valuable tests and techniques. Some of these, of more interest to geologists, are simply mentioned here; for details, go to the texts, e.g. Miller and Kahn, Snedecor, etc.

Much geologic data can be presented in the form of scatter plots, wherein we wish to see how one property is related to another. Examples are plots of roundness versus distance; mean grain size versus sorting; feldspar percentage versus stratigraphic position; for a collection of dinosaur bones, length of thigh bone versus thickness of the bones; zircon/tourmaline ratio versus grain size; percent carbonate mud versus roundness of shell fragments, etc. To analyze such associations, two main procedures are applied: (1) the perfection of the association is tested, and (2) the equation of the relationship is determined.

If the two properties are very closely related, they give a long narrow "train" of points on a scatter diagram. If the two properties are not associated, a random "buckshot" pattern emerges. The correlation coefficient, r, computes the perfection of correlation. For perfect correlation r = 1.00, which means that, knowing one property, we can predict the other property exactly, and that both increase together. An r of -1.00 means perfect negative correlation, a correlation just as exact except that as one property increases the other decreases. If the two properties are not correlated, r may be .00; weak correlation would be +.25 or -.15, etc. Coefficients beyond $\pm .50$ are considered "good" for most geological work. The normal correlation coefficient is valid only for straightline trends. Other methods must be used for hyperbolic, parabolic, sinusoidal, etc., trends.

If a small number of data points are available, it is possible for "good-looking" correlations to arise purely by chance. Thus one should always refer to tables which show whether the given value of r shows a significant correlation; this depends of course on the number of samples and the value of r, thus is similar in principle to the t test.

Squaring the correlation coefficient, r, gives the coefficient of causation r^2; this tells one how much of the variation in one property is explained by the variation in the other property. For example, if we find that in a series of pebbles, the roundness shows a correlation coefficient of r = +.60 with grain size, then $r^2 = .36$, and we can say that 36% of the variation in roundness is caused by changes in grain size (thus 64% of the roundness variation would be due to other causes: differences in lithology, distances of travel, "chance", etc.). Further analyses may be carried out, such as partial or multiple correlations, analysis of variance, etc.--see standard texts.

A trend line may be fitted to a scatter diagram, and an equation may be fitted to this line so that, given a value of one property, the other property may be predicted. Trend lines can be drawn in by eye, but this process is usually sneered at; a mathematical way of doing it is the "least squares" method. It is important to realize that, unless the two properties are almost perfectly correlated, two trends lines can be computed; let's say we have a graph of feldspar content versus roundness, with a correlation coefficient of +.45 (good, but far from perfect correlation). We can either (1) compute the line to predict most accurately the roundness, given the feldspar content; or (2) a line to predict most accurately the feldspar content given the roundness. The two lines may easily form a cross with as much as 30° or more difference in angle. Again, these work only for straight-line trends; non-linear trends require more complicated arithmetic.

To the trend line is usually attached a "standard error of estimate" band, essentially equal to the standard deviation. This band runs parallel to the computed trend and includes two-thirds of all the points in the scatter diagram. Its purpose is to show the accuracy of the relationship. For example, in a certain brachiopod the length and width of the shell are related by the equation L = 2W -1 ± 0.5. The last figure is the standard error of estimate; if a given specimen has a width of 3.5 cm, the length is most likely 8 cm, but we can expect two-thirds of the specimens to range between 7.5 and 8.5 cm. (one-sixth of them will be over 8.5 cm, and one-sixth under 7.5cm).

Abridged Table of t

This table can be used only when the data is in the form of means of continuous variables.

Degrees of Freedom	P						
	0.50	0.20	0.10	0.05	0.02	0.01	0.001
1	1.0	3.0	6.3	12.7	31.8	63.7	636.6
2	0.84	1.89	2.92	4.30	6.97	9.93	31.60
3	0.79	1.62	2.35	3.18	4.54	5.84	12.94
4	0.78	1.52	2.13	2.78	3.75	4.60	8.61
7	0.73	1.42	1.90	2.37	3.00	3.50	5.41
10	0.70	1.36	1.81	2.23	2.76	3.17	4.59
20	0.69	1.30	1.73	2.09	2.53	2.85	3.85
30 - ∞	0.68	1.28	1.65	1.96	2.33	2.58	3.29

Enter the table with the proper degrees of freedom and read right until you reach the (interpolated) value of t you obtained by calculation. Then read up to the top of the table the corresponding P. Example: for 10 d.f., t = 3.0; therefore P -about .013, i.e., there is a little more than 1 chance in 100 that the differences are due to chance. As a general rule, if P is .05 or less, the differences are considered as real: if P is between .05 and .20, there may be real differences present, and further investigations are warranted with the collection of more samples if possible; if P is over .20 differences are insignificant.

$$t = \frac{\bar{x}_a - \bar{x}_b}{s} \sqrt{\frac{n_a \cdot n_b}{n_a + n_b}}$$

Extended values for t for d.f>30, based on Fisher and Yates, Statistical Tables, with linear extrapolations,

$P = 10^{-4}$, $t = 3.9$	$P = 10^{-12}$, $t = 7.2$ (trillion)
$= 10^{-5}$, $t = 4.4$	$= 10^{-15}$, $t = 8.2$ (quadrillion)
$= 10^{-6}$, $t = 4.9$ (million)	$= 10^{-18}$, $t = 9.3$ (quintillion)
$= 10^{-7}$, $t = 5.3$	$= 10^{-21}$, $t = 10.3$ (sextillion)
$= 10^{-8}$, $t = 5.7$	$= 10^{-24}$, $t = 11.3$ (septillion)
$= 10^{-9}$, $t = 6.1$ (billion)	$= 10^{-27}$, $t = 12.3$ (octillion)

Abridged Table of X^2

<u>This table can be used only when the data is in the form of discrete counts of individuals.</u>

Degrees of Freedom	P						
	0.50	0.20	0.10	0.05	0.02	0.01	0.001
1	0.45	1.65	2.71	3.84	5.41	6.64	10.83
2	1.39	3.26	4.61	5.99	7.82	9.21	13.82
3	2.37	4.65	6.25	7.82	9.84	11.3	16.2
4	3.36	5.91	7.7	9.4	11.6	13.2	18.4
6	5.35	8.60	10.6	12.5	15.0'	16.8	22.4
8	7.34	11.0	13.3	15.5	18.1	20.0	26.1
10	9.34	13.4	15.9	18.3	21.1	23.2	29.5
15	14.3	19.3	22.3	25.0	28.2	30.5	37.7
20	19.3	25.1	28.4	31.4	35.0	37.5	45.3
25	24.3	30.7	34.3	37.6	41.5	44.3	52.6
30	29.3	36.2	40.2	43.7	47.9	50.8	59.7

Enter the table with the proper degrees of freedom and read right until you reach the (interpolated) value of X^2 you obtained by calculation. Then read up to the top of the table the corresponding P. Example: for 8 d.f., $X^2 = 14.1$; therefore P = about .08, i.e., there are 8 chances in 100 that the differences are to chance. As a general rule, if

P is .05 or less the differences are considered as real; if P is between .05 and .20, there may be real differences present and further investigations are warranted with the collection of more samples, if possible; if P is over, .20 differences are insignificant.

$$X^2 = \Sigma\left(\frac{D^2}{E}\right)$$

Table of Areas of the Normal Probability Curve

The following table gives the percentage of values included within a range of the mean plus or minus the number of standard deviations listed in the left-hand column. More complete tables can be found in textbooks.

Number of Standard Deviations d	Percentage of Samples Included in Range $\bar{x} \pm d$	Percentages of Samples Outside the Range $\bar{x} \pm d$
0.1	8%	92%
0.2	16%	84%
0.3	24%	76%
0.4	31%	69%
0.5	38%	62%
0.6	45%	55%
0.68	50%	50%
0.7	52%	48%
0.8	58%	42%
0.9	63%	37%
1.0	68.3%	31.7%
1.2	77%	23%
1.4	84%	16%
1.6	89%	11%
1.8	93%	7%
2.0	95.5%	4.5%
2.2	97.22%	2.78%
2.4	98.36%	1.64%
2.6	99.06%	0.94%
2.8	99.48%	0.52%
3.0	99.73%	0.27%
3.5	99.954%	0.046%
4.0	99.9947%	0.0063%
5.0	99.999943%	0.000057%
6.0	99.9999998%	0.00000020%

The table is used in solving the following types of problems:

(1) Chert pebbles on a beach have a mean sphericity of 0.71, standard deviation of 0.08. What percentage of the pebbles will have sphericities between 0.68 and 0.74, assuming the distribution is normal? The stated limits are 0.03 on either side of the mean; .03 is in this example 0.375 standard deviations, and approximately 29% of the pebbles will fall within this range. (b) Out of 500 pebbles, how many will be expected to have sphericities higher than .90? This value, .90 is .19 higher than the mean, i.e., 2.375 standard deviations; 2% of samples fall outside the range of $\bar{x} \pm 2.375\ \sigma$, and of these half will be above, half below; therefore the answer is 1%, or 5 pebbles.

(2) A coin weighs an average of 15.0 gm as minted. Due to irregularities in the minting process, the standard deviation is 0.6 gm. Ten percent of the coins will be heavier than what weight? The table is constructed to give us the percentage of samples falling outside the given range, both above and below; we need only the ones higher, therefore we look in column 2 for twice 10%, or 20%; the corresponding value is 1.3 standard deviations. In other words, 20% of the samples will fall outside the range of $\bar{x} \pm 1.3\sigma$, 10% being below and 10% above. 1.3 standard deviations in this case is about 0.8 gm; therefore the answer is 15.8 gm.

MINERAL COMPOSITION OF SEDIMENTARY ROCKS

In igneous rocks, many minerals occur commonly and are important rock-formers. In sedimentary rocks, however, only a few minerals are important rock-formers, but those few minerals may be present in a bewildering number of varieties. For example in one piece of limestone the mineral calcite may be present as oolites, reworked pebbles, fossils, pellets, microcrystalline ooze, several generations of pore-filling cement, and veinlets. Furthermore, in normal igneous rocks essentially all the minerals are direct chemical precipitates with no transported or abraded particles and little replacement; but in sediments one must distinguish between (1) terrigenous minerals, and (2) chemical minerals, divided into (2a) allochemical minerals (transported) and (2b) orthochemical minerals (forming in place). Orthochemical minerals furthermore may be formed either by replacement of previous minerals or by direct precipitation. A given mineral species may be present in all of these categories even within the same rock.

The abundance of a terrigenous mineral in a sedimentary rock depends on three factors (Krynine):

(1) Availability. The mineral must be present in sufficient abundance in the source area. One cannot get an arkose by eroding limestone, or chert pebbles by eroding granite. Similarly lack of feldspar may be due not to a humid climate but to the fact that the source was an older sandstone, or a low-rank phylite or schist.

(2) Mechanical durability, favored by lack of or poor cleavage, and high hardness. Long abrasion selectively eliminates soft or easily cleaved minerals.

(3) Chemical stability. Minerals that form late in igneous rocks, when those bodies are crystallizing under cooler and more hydrous conditions, are most stable in sediments because they are more nearly adjusted to the relatively cold and wet sedimentary environment. The order of chemical stability is approximately the reverse of Bowen's reaction series, but local chemical conditions can alter the order (e.g. Todd, 1968 JSP).

Stability Series for Terrigenous Minerals

Quartz, Zircon, Tourmaline
Chert
Muscovite
Microcline
Orthoclase
Albite
↕

Hornblende, Biotite
Pyroxene
Anorthite
Olivine

Minerals above the dashed line can also form by chemical precipitation in sediments. When they do form in place, they are often termed "authigenic."

RELATIVE ABUNDANCE OF MINERALS IN THE SEDIMENTARY ROCKS

TERRIGENOUS MINERALS, derived from erosion of source lands; comprise 60% to 80% of the stratigraphic column.

QUARTZ, **35-50%

CLAY MINERALS 25-35% (Authigenic alteration common). Main types:
- "Sericite"* (impure fine grained muscovite, HKAlSiO).
- Illite group* (HKAlSiO)
- Montmorillonite group* (HMgAlSiO)
- Chlorite group* (HMgFeAlSiO with Si low, FeMg high).
- Kaolin group* (HAlSiO)
- Bauxite group* (HAlO) (not strictly a clay)

METAMORPHIC ROCK FRAGMENTS 5-15%
(not a mineral; includes detrital particles of slate, phyllite, schists, metaquartzite)

FELDSPAR** 5-15%
(K feldspar considerably more abundant than plagioclase)

CHERT** 1-4%
(mostly reworked from older cherty limestones)

COARSE MICAS* 0.1-0.4%
(Muscovite HKAlSiO by far most common; some Biotite HKMgFeAlSiO and Chlorite HMgFeAlSiO)

CARBONATES 1.2-1%
(reworked fragments of older limestones)

ACCESSORY (HEAVY) MINERALS 0.1-1.0%
- Opaques**:
  - Magnetite, Ilmenite, Hematite, Limonite and Leucoxene
- Ultrastable group
  - Zircon**, Tourmaline, Rutile
- Less stable group:
  - Garnet, Apatite, Kyanite, Staurolite, Epidote, Hornblende, Pyroxene and about 100 others of minor importance volumetrically.

CHEMICAL MINERALS, precipitated from solution within the basin of deposition; comprise 20% to 40% of the stratigraphic column.

CARBONATES,** 70%-85%
Calcite 2/3 to 3/4; dolomite 1/3 to 1/4; aragonite, siderite, ankerite minor

SILICA 10-15%
(mostly quartz and chert; a little opal)

SULFATES AND SALTS** 2-7%
(Chiefly gypsum, anhydrite, halite; some sylvite and other K salts; a little barite, etc.)

MISCELLANEOUS 2-7%
(Feldspar, hematite, limonite, leucoxene, pyrite,** phosphates, glauconite, manganese, tourmaline, zircon, rutile, anatase, zeolites and a good many others).

You are required to know the exact chemical formulae of the minerals marked **, and of all the minerals named under the groups marked **. Know the chief elements listed for each mineral marked *. A good knowledge of the chemistry of earth minerals is vital to understanding their behavior in the sedimentary environment.

Quartz

Composition, SiO_2, essentially constant with no isomorphous substitution. Hexagonal (rhombohedral). Hardness 7, cleavage for practical purposes lacking N_o 1.544, N_e 1.553 (both almost constant), both indices slightly higher than Balsam.

Bn 0.009 (quite constant), giving gray to white or cream colors. Extinction in crystals parallel (or symmetrical to rhomb faces), lengthslow. Often shows undulose extinction or "strain shadows", usually as the result of stress after crystallization, occasionally the result of the method of growth (even unstrained quartz crystals in geodes and veins sometimes show undulose extinction--partially this may represent a transition to chalcedony).

Strain shadows usually ocur in bands subparallel to the c axis, and in strongly strained grains the c axes may deviate as much as 30° (true angle). Intense strain may result in the formation of warped, subparallel lines made of minute bubbles (Böhm lamellae). Quartz is uniaxial positive, but becomes slightly biaxial when strained, with 2V of 5° or more. Please get a few interference figures on quartz grains showing higher interference colors (pale yellow), so that you will become familiar with the quartz flash figure and not confuse it with the truly biaxial feldspar figure. Luminescence Petrography (Sippel) reveals much occult strain, fracturing and growth zones.

The vast majority of quartz grains contain inclusions (see publications by Roedder). Most commonly a few isolated vacuoles or trains of vacuoles are found; these are usually filled with liquid alone, but often include a tiny spherical, "dancing" gas bubble, and rarely are filled with gas alone. Liquid-filled vacuoles appear brownish (a complex dispersion effect), and have a high relief with index below quartz; in reflected light they appear silvery. Gas-filled bubbles have such a tremendous negative relief (1.00 vs. 1.544) that they appear black and opaque; but they are silvery in reflected light. Sometimes the vacuoles have the external shape of a quartz crystal ("negative crystals"), but usually they are highly irregular with rounded protuberances. Most quartz grains have only a few vacuoles or bubble trains; some, however, are so crowded with bubbles that they appear milky in reflected light. These excessively bubbly types come from hydrothermal veins. Mineral inclusions are fairly common; tourmaline, mica, magnetite, hair-like rutile needles (sometimes in sets intersecting at 60° angles), chlorite, feldspar or zircon. A zonal arrangement of fluid or mineral inclusions, parallel with quartz crystal faces, indicates that the quartz grew in an open cavity as a vein or geode filling.

Availability. Quartz forms 35-50% of the terrigenous fraction of sedimentary rocks. Most quartz, especially that in Mesozoic and later rocks comes from reworking of older sandstones or sandy limestones. This can be recognized by anomalous size-roundness relations, high roundness standard deviation within the same grain size (i.e., association of well-rounded with angular grains of the same size), or more rarely presence of abraded quartz overgrowths. See Syllabus on Grain Shape for details. Also, "side evidence" may be used such as the presence within the same specimen of considerable chert, shale or limestone fragments, or the character of the heavy minerals. A considerable amount of quartz comes from metaquartzites or schists, and a small amount from hydrothermal veins or volcanic material; however, the ultimate source of most quartz is to be found in the granites and granite-gneisses. Criteria are given below for distinguishing these types.

Durability. Quartz is the most durable of the abundant minerals because of its hardness, toughness, and lack of cleavage. To round a quartz grain takes a tremendous amount of time; thus quartz is usually subangular in sediments (see section on shape). Grains are usually subequant, but those derived from schists are usually slightly more elongated or platy. In silts it is fairly common to find the grains elongated parallel to the c axis, probably the result of obscure prismatic and rhombohedral fracture. Terrigenous quartz grains as fine as 3 microns or even less occur, but .010-.020 mm is about the practical lower limit of quartz in most mudrocks. Fractured quartz breaks easily on transport (Moss, '72 JSP), as does poly-crystalline (Harrell and Blatt '78 JSP).

Stability. Quartz is ultra-stable under nearly all surface conditions, and it is probable that very little quartz is ever dissolved by weathering; but some solution of quartz in tropical areas has been reported with corrosion of grains (Crook, 1968) and corrosion (by organics) can even occur in soils of S.E. USA (Cleary and Conolly '72 JSP), producing skeletal grains by solution along fractures (microgriking). Folk ('78 JSP) reviews quartz solution. However, on burial in the sub-surface, even under only moderate pressures and temperatures, quartz may occasionally dissolve; thus, some sandstones show sutured quartz grains along styolites, and even styolites in limestone often cut through and dissolve quartz grains. Dolomite or calcite rarely replace the edges of quartz grains, to form etched or pitted contacts (quartz in limestones or dolomites is often thought to be replaced when it is really just an optical illusion caused by "overlap" of carbonate around the curving edge of the quartz grain). It can also be replaced by pyrite. On the other hand, quartz grains commonly have authigenic quartz overgrowths, showing that the mineral may either grow or be destroyed depending on conditons.

Depositional Characteristics. Quartz in sandstones is uniformly distributed, in mudrocks usually occurs as laminae or clots, and in limestones may occur as laminae or as scattered grains or nests. Grains often show physical orientation parallel with the bedding, in addition, they also more rarely show an optical orientation with the slow ray (c axis) parallel with the bedding. Check this by inserting the gypsum plate; a dominance of blue grains in one direction and yellow grains in a direction of 90° indicates optical orientation. This is partially due to differential abrasion, with the "c" axis direction being harder so the grains tend to become elongated parallel to "c", and partially due to a weak tendency to fracture into chips parallel to "c" rather than perpendicular to it.

Quartz Varieties. The study of quartz types is one of the most fascinating and valuable aspects of sedimentary petrography. It is of great value in paleogeographic interpretation, and is increasingly used in the correlation of formations. H.C. Sorby (1858, 1877, 1880) first studied inclusions and extinction in quartz grains as a clue to their source. Later work along this line was done by Mackie in the 1890's, and the subject achieved great height under Krynine in the 1940's. Blatt (1960's) has made an intensive quantitative study of quartz in source rocks and sediments, and Basu, Suttner and others have made further advances in the '70's. Quartz is an ideal mineral to use because there are so many possible varieties, and because many of them can be assigned to a definite type of source area. Each environment of quartz formation--batholith, volcano, schist, pegmatite, fault zone--lends its characteristic impress to the properties of the quartz grains, and these properties remain when the grain is deposited in sediments derived from these areas, as first shown by Sorby.

There are two systems of classification of quartz types. In the genetic classification, developed by P.D. Krynine, an attempt is made to allot each quartz grain to a specific environment of formation: plutonic, volcanic, schistose, stretched metaquartzite, recrystallized metaquartzite, hydrothermal, etc., based on type of

ESTIMATED PROVENANCE OF QUARTZ

Intended for Sand – Sized Grains

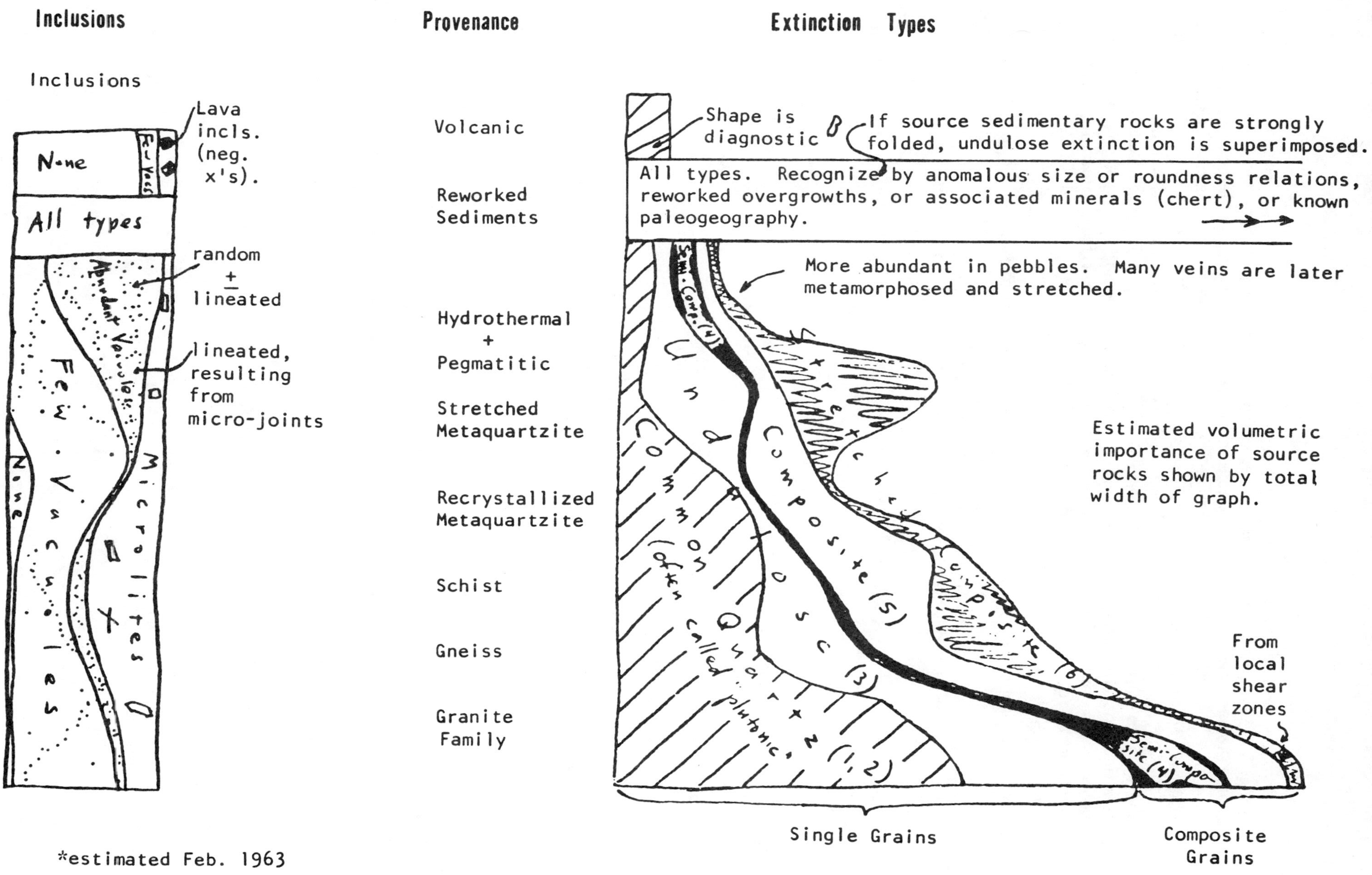

*estimated Feb. 1963

extinction, inclusions, and grain shape. This is always highly subjective, often difficult and requires considerable experience. However it is the ultimate aim of the study of quartz types and the most valuable system of classification because through it the character of the source area can be determined, and much valuable paleogeographic interpretation can be made. In the empirical classification, a binomial series of pigeonholes is set up based on 6 different types of extinction and 4 types of inclusions, giving a total of 24 arbitrary classes. These classes tie up approximately with the genetic classification, but there is considerable overlapping (for example, straight extinction quartz with few bubble inclusions may come from plutonic, recrystallized metaquartzite, or vein environments). This is an objective classification, is easily learned and hence, is amendable to statistical techniques and the comparison of sets of samples by means of the X^2 test. Statistical tests are difficult to apply to the genetic classification because of its subjectivity. The best procedure is to count a hundred or so grains using the empirical classification, then based on this, and with the genetic classification in mind, conclude about the provenance of the grains.

Genetic Classification of Quartz Types
(After P. D. Krynine)

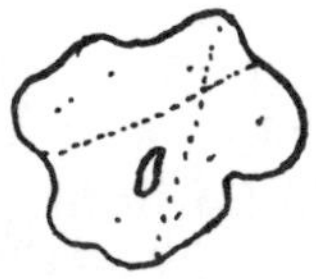

COMMON (PLUTONIC)

1. Straight to slightly undulose extinction.
2. Some vacuoles; may be a few microlites
3. Subequant, xenomorphic
4. Typical of granites; but also furnished by many other sources.

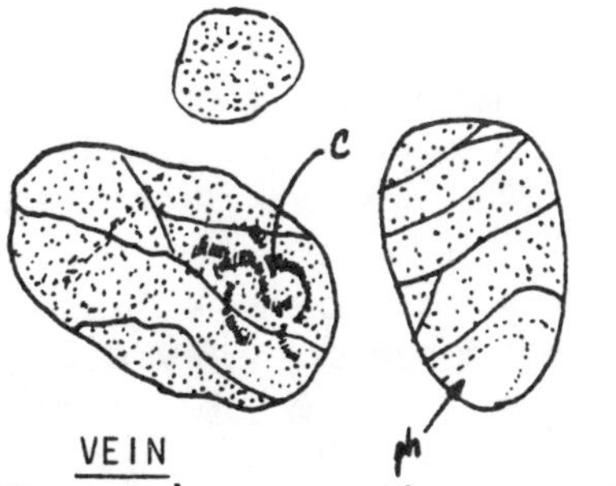

VEIN

1. Abundant vacuoles; sometimes vermicular chlorite (c); rarely, zoned phantom crystals (ph).
2. Semi-composite to straight or undulose extinction, some may be badly sheared. Often shows comb structure.
3. Often forms large grains and pebbles.

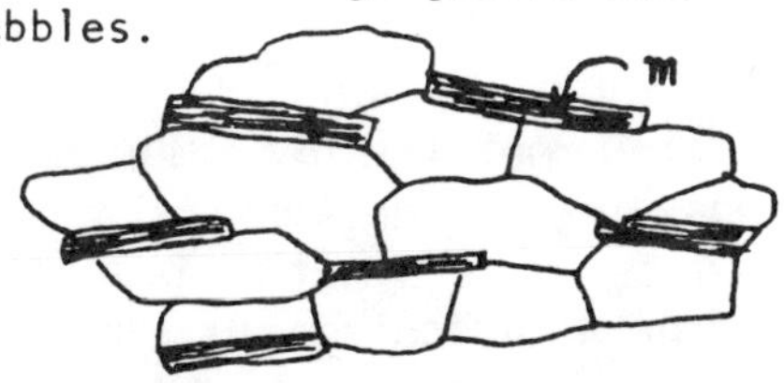

SCHISTOSE METAMORPHIC

1. Elongate, composite with straight borders.
2. Mica inclusions (m).
3. Straight to slightly undulose extinction.

Perfect crystal in 3-D view. No prism faces.

VOLCANIC

1. Idiomorphic hexagonal-bipyramidal shape with perfectly straight sides and rounded corners. Prism faces very small or lacking. Embayments common.
2. Almost no inclusions; water-clear. May have negative crystals (n).
3. Straight extinction.

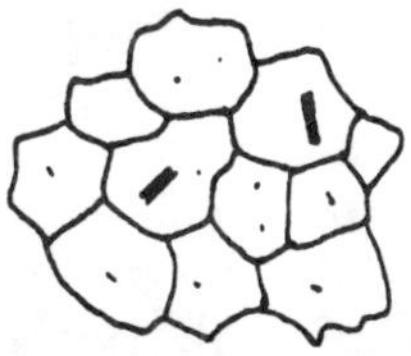

RECRYSTALLIZED METAMORPHIC

1. Straight boundaries between equant interlocking grains, forming mosaic.
2. May be some microlites & vacuoles.
3. Straight to slightly undulose extinction.

(Do not confuse with quartz-cemented sands!)

STRETCHED METAMORPHIC

1. Strong undulose extinction; borders may be smooth (sm), crenulated (cr) or granulated (gr).
2. Elongate, lenticular shape of crystal units.
3. Some microlites and vacuoles.

Genetic Classification (modified from Krynine)

I. Igneous Quartz

A. Plutonic Quartz, from granite batholiths or granite-gneisses, thus the most abundant type. Shape xenomorphic, irregular subequant, sometimes with re-entrant angles. Single crystals average about 0.5 mm. Often has straight to slightly undulose extinction in sand-size grains, but may be strongly undulose if the batholith has been stressed after solidification, or if the rock is gneissic. A few grains may be composites. Usually contains no inclusions other than a small amount of scattered vacuoles; if these are randomly scattered they were probably trapped during crystallization, but if arranged in lines (actually planes) they probably formed upon stressing of the parent rock and represent incipient, healed fractures. Mineral inclusions sometimes occur but are not common; rutile, zircon, micas, feldspar, biotite, hornblende, tourmaline, etc. are most frequent. Because much quartz from other environments (metamorphic, vein, etc.) has the same characteristics as "plutonic" quartz, a better name for this type is "common" quartz.

B. Volcanic Quartz, as phenocrysts derived from erosion of volcanic rocks such as rhyolites, latites, dacites, etc. or as direct contributions from ash falls, is not abundant in general but some stratigraphic sequences contain numerous grains of it, and it is an excellent stratigraphic marker. Recognition is based mainly on shape--whole or fragmental bipyramidal crystals lacking prism faces, hexagonal with geometrically straight edges (compare with hairline microscope), but usually with rounded corners. They often have large rounded corrosion embayments. Extinction nearly always straight, usually contain no inclusions and thus are water-clear. There are rarely any bubble inclusions but occasionally bits of volcanic glass or aphanitic volcanic ground-mass may be trapped, sometimes in the form of negative hexagonal crystals. On abrasion, the diagnostic features of volcanic quartz are lost. Be on the conservative side when identifying quartz as volcanic. In recent sands, volcanic phenocrysts are easily recognizable by their highly polished surfaces and water-clear appearance, together with the typical hexagonal bipyramid shape and embayments.

C. Vein Quartz, derived from pegmatites, hydrothermal and much more rarely sedimentary vein fillings. It is very common but not usually abundant except in certain stratigraphic horizons; it is possibly the most comon type of quartz occurring in pebbles. Shape is not diagnostic. Extinction is most commonly semi-composite with parallel arrangement of individuals almost but not quite in optical continuity, partly a reflection of the comb structure produced as the crystals grow out all nearly perpendicular to the walls of veins. Many grains have straight extinction, however, and sometimes the extinction may be slightly undulose, because the formation of veins is often accompanied by stresses; intense stress and shearing causes vein quartz in some cases to take on a metamorphic appearance with intense strain shadows and crenulations, but this is then arbitrarily classed as stretched metamorphic quartz. Grains contain no microlites as a rule (occasionally mica or tourmaline or feldspar may occur if it is from a pegmatite, or carbonate if a cooler vein) but in most cases vacuoles are very abundant, giving the typical milky appearance to pebbles and even sand grains, because the mineral formed in a very hydrous environment. However, some grains contain few vacuoles. Vermicular chlorite is very diagnostic of hydrothermal quartz, and the presence of zoned inclusions (parallel to old crystal faces as the crystal grew) is also diagnostic.

Any type of quartz grain that has been slightly deformed may readjust by developing a mosaic of almost parallel-oriented sub-grains; this process is analogous to the process of "polygonization" in strained metals, which also results in semi-composite extinction mimicking the appearance of a comb-structured vein. Quartz should not be referred to as "polygonized" unless the formative process is known: "semi-composite" is better as a purely descriptive, non-genetic term.

II. Metamorphic Quartz

A. Recrystallized Metamorphic Quartz. Formed under intense but non-shearing stresses when quartz recrystallizes, or else by recrystallization after stress has passed. In the lab, recrystallization occurs at 600-800°C (Tullis, '73 GSA). Occurs in many recrystallized metaquartzites, highly metamorphosed and gneissic rocks. Shape subequant. Extinction is straight (recrystallization destroys any previous strain), and grains may be either single individuals or composites made up of a mosaic of equidimensional grains with rather straight boundaries, and widely different optic orientation. Contains few inclusions, mostly vacuoles; appears to have less vacuoles than plutonic quartz. May have mineral inclusions in small amounts, feldspar, micas, tourmaline, etc. Usually it is quite difficult to tell from plutonic quartz (often impossible), but despite its lack of recognition probably is an important contributor to sands. Caution: the chief diagnostic feature is the presence of composite grains; naturally these will be less evident if the sandstone is finer, so in comparing samples you should work with ones of about the same grain size; also on prolonged abrasion or chemical weathering composite grains will break apart into their sub-individuals and look like single grains of plutonic quartz.

B. Schistose Quartz forms during lit-par-lit injection of schists, also on recrystallization of schistose rocks. Quartz grows in between parallel mica flakes, hence is often elongated or platy, with sub-parallel straight edges; since it forms by recrystallization or growth from solution after stress has passed, it seldom has strain shadows and usually shows straight extinction. Composite grains are common. It sometimes has no inclusions and vacuoles are uncommon, but micaceous inclusions or other metamorphic minerals are frequent. It is a common type of quartz but difficult to recognize if there has been prolonged abrasion, because the chief diagnostic feature is the shape. Its presence can be shown by a statistical count of the length-width ratios of grains in a slide (Bokman), an idea first proposed by Sorby.

C. Stretched Metamorphic Quartz (or Sheared Quartz) is very easy to recognize and is quite important in sediments. It forms when a quartz-bearing rock (be it older sandstone, granite, schist, or quartz vein) is sheared or strained in the absence of recrystallization. Grains are generally elongated or platy. Extinction is moderately to strongly undulose, and grains may be single but are usually made up of sub-individuals of lensoid or "smeared-out" shape. Often boundaries between sub-individuals are intensely crenulated or sutured, sometimes finely granulated. Sometimes metamorphic inclusions--

micas, sillimanite, garnet, etc,--are present. This type of quartz is common in graywackes. Cautions: (1) do not confuse fine-grained stretched metamorphic quartz with chert. Microcrystals of chert seldom are over 10 microns, and are randomly oriented as a rule; the component individuals of stretched metamorphic quartz are usually elongated, coarser than 20 microns, and are in sub-parallel orientation (test with gypsum plate). (2) On prolonged abrasion metamorphic quartz may possibly abrade or fracture faster than plutonic quartz, so in very well-rounded sands the percentage of metaquartz may be abnormally low. (3) Be careful not to be confused by post-depositional strain-shadows when studying quartz types; sandstones made entirely of straight-extinction quartz develop strain shadows where strongly folded or faulted, with most strain occurring where a grain is "pinched" by corners of a neighboring detrital grain.

III. Reworked Sedimentary Quartz.

Sedimentary authigenic quartz is sometimes found as reworked detrital grains. The most common kind is as quartz grains with reworked overgrowths, formed in an older sandstone or limestone. Overgrowths survive river transport, and probably most beach transport, with very little sign of abrasion; so do not try to identify a reworked overgrowth by its so-called "abrasion", because this cannot be seen in thin section (unless the grain is broken right slap in two). If, in a sandstone, scattered grains have good overgrowths and others have none, and particularly if the nuclei of the overgrowth grains have a uniform shape (e.g. all are well rounded), then they are probably reworked. Indigenous overgrowths usually interlock with long or sinuous contacts, and occur on most of the grains in the slide regardless of the shape of the nucleus. Occasionally quartz nodules are reworked from older carbonate rocks, where they occur associated with chert nodules. These consist of flamboyant quartz and are quite scarce; they often originate through replacement of evaporites.

Empirical Classification (Folk). Six extinction types are recognized for counting purposes; (1) Straight, the grain extinguishes uniformly all over at once. (2) Slightly undulose, the grain is a single individual and the extinction shadow sweeps smoothly and without breaks across the grain; however the sweep is accomplished on very slight rotation of the stage. (3) Strongly undulose, the grain is a single individual and the extinction shadow sweeps smoothly across, but a large rotation of the stage is required betwen the time one part of the grain is extinguished to the time another part is extinguished. (It is difficult to establish objective boundaries between the three classes, but experience has shown that in grains of .15-.35mm, "straight" grains extinguish with a stage rotation of less than 1°, "slightly undulose" grains require a rotation of 1° to 5° before the deepest part of the extinction shadow sweeps from one side of the grain to the other, and "strongly undulose" grains require a rotation of over 5°, often as much as 20° or more. Blatt says that this measurement is useless if made on a flat stage because we do not see the true extinction angle). (4) Semicomposite, the grain is made up of two or more sub-individuals with very close optical orientation, but there is a distinct break between individuals and the extinction shadow does not sweep smoothly across; the individuals may however have straight to slightly undulose extinction. The presence of crenulated borders or strong undulose extinction throws this into class "6." (5) Composite, the grain in made up of two or more sub-individuals, with widely differing orientation, but these same individuals have normal (not crenulated) boundaries and straight to slightly undulose extinction. (6) Composite metamorphic, the grain is made up of two or more sub-individuals with strongly undulose extinction, and may or may not have crenulated boundaries.

There is of course some degree of tie-up between the genetic and empirical classifications. As shown on page 71, types 1 and 2 come mostly from granites and gneisses, to a smaller extent from recrystallized metamorphic rocks, and to some extent from veins. Types 3 and 6 come chiefly from gneisses and stretched metamorphic sources, but type 3 is in some strained granites. Type 4 is typical of hydrothermal veins (but also occurs in lesser amounts in other rocks), and type 5 occurs chiefly in schists or recrystallized metamorphic loci, but to a small extent in granites.

The second basis for the empirical classification is the type of inclusion present. Here some leeway is present and for each problem a different classification might be set up; for example, in some areas vermicular chlorite inclusions might be diagnostic and would be set up as a separate class for counting; or, the presence of negative crystals or of tourmaline inclusions might be of value and would be counted separately. For general purposes, the following classification is suggested, operated under the principle that if a grain does not fit in the first category, one goes on down the line until he finds the first category that applies (for example, a grain with abundant bubbles and rutile needles would be placed in class "a"). The types are as follows: (a) Abundant vacuoles (grain may even appear milky in reflected light; (b) Rutile needles or other hairlike inclusions; (c) Microlites (inclusions of minerals other than rutile); (d) Sparse vacuoles, no microlites. For detailed work, microlites should be identified and counted separately. Of these classes, type "a" is diagnostic of hydrothermal veins; "b" is an excellent stratigraphic and regional marker, but its genetic significance is characteristic of metamorphic terranes, but also occurs in plutonic environments; and "d" is the commonest type, derived from all sources.

Empirical Classification of Quartz Types
(After R. L. Folk)

A. BY EXTINCTION (ALL SHOWN AS UNDER CROSSED NICOLS)

1. Single grain, straight extinction.

2. Single grain, slightly undulose extinction.

3. Single grain, strongly undulose extinction.

4. Semicomposite grain, straight to slightly undulose extinction.

5. Composite grain, straight to slightly undulose extinction.

6. Composite grain, strongly undulose extinction.

B. BY INCLUSIONS

a. Abundant vacuoles

b. Rutile needles

c. Microlites

d. Few vacuoles no microlites

In practice, one examines a series of slides to set up the diagnostic quartz types present, patterning it after the above classes. If necessary he erects additional classes to take care of types he thinks might be significant. Then he proceeds to count 100 or more grains (always under high power), and classifies each grain as "1b", "3a", etc., recording them in a binary table. If special types, such as volcanic quartz or reworked abraded grains are present, they may be counted in a separate class. After this has been done, he re-examines the slide with the genetic classification in mind, refers to the graph, and decides the probable lithology of the source area. To do this he also takes into account the associated minerals, and the effects of grain size and abrasion on his count (less composite grains will be found if the material is fine-grained, etc.). Criteria for an older sedimentary source have already been indicated. In general grains from a sedimentary source area are of diverse types; if quartztypes are few the source is probably a primary igneous rock.

An Opposing View on Quartz Extinction. H. Blatt (1963-1967) has recently made study of quartz in igneous and metamorphic rocks, in the immediate disintegration products of such rocks (grus), and in consolidated sedimentary rocks from many areas. His results do not support the "classical" interpretation given above. (1) Blatt measured the true degree of undulatory extinction in 586 medium sand-sized quartz grains from freshly disintegrated granites, schists, and gneisses. He found (a) that 91% of these grains have less than seven degrees of undulose extinction; (b) there is no correlation between rock type and amount of strain of the grains; (c) no grain had undulose extinction greater than eighteen degrees. From these data, Blatt concluded the strength of the quartz structure does not generally allow the development of more than ten degrees of undulose extinction. This result is consistent with the results of experimental deformation of quartz. The important point is that Blatt claims that no real difference exists in degree of undulose extinction in quartz grains among igneous and metamorphic rocks, and therefore that any "averaging" method of determining amount of undulose extinction on a flat stage is at least as useless as attempts at differentiation with universal-stage measurements. (2) Blatt further says that if one defines "strongly" undulatory as greater than five degrees of undulose extinction, as Folk does, there are very few truly strongly undulatory grains, and most grains that appear to be "strongly" undulatory appear that way only as a result of the relationship between the plane of the thin-section and the plane containing the optic axes in a strained grain. (3) Blatt's data indicates that, based on universal-stage measurements, the percentage of grains with greater than one degree of undulose extinction (Folk's slightly and strongly undulatory grains) is statistically identical among granites, schists, and gneisses. Therefore, there appears to be no way of distinguishing between quartz grains from each of these three rock groups on the basis of either the unstrained/-strained ratio or on the basis of the slightly strained/strongly strained ratio. (4) Blatt feels that medium sized polycrystalline quartz grains composed of ten or more sub-grains are an excellent indicator of metamorphic source (composite quartz of Folk). However, he considers that medium sized grains composed of less than five sub-grains are as commonly derived from granites as from schists and gneisses. (5) Blatt's data shows that well-rounded, pure quartz sandstones (i.e. supermature orthoquartzites) have a very high percentage of "straight" quartz, much higher than any conceivable source rock. He interprets this as being due to selective destruction of undulatory quartz, while the unstrained quartz survives. This had been suspected for some time by many workers. Basu et al. (1975 JSP) and Young (1976 JSP) have come to a practical middle ground in a detailed re-evaluation of undulosity and polycrystallinity.

A key point to the Cambro-Ordovician sandstones of the central U.S., which consist almost entirely of well-rounded common (straight to slightly undulose) quartz;

The Quartz Continuum

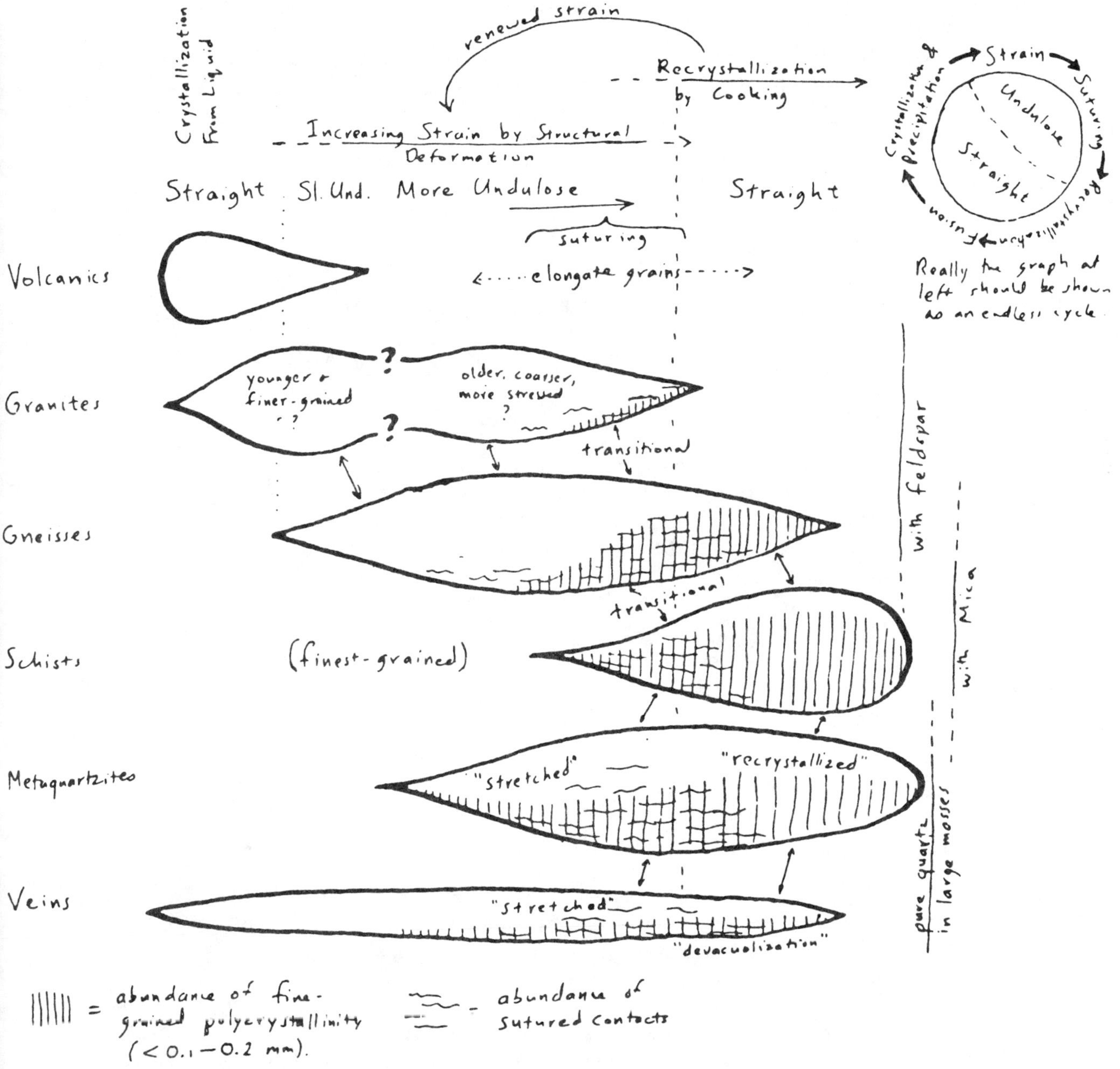

Note: Two geologic realms where straight extinction occurs, at opposite ends of the scale: (1) crystallization from magma or solution, (2) recrystallization of a solid. In between is the realm of undulose quartz. Undulosity is a function of size as well as strain. Thus schists have so much straight quartz because they are the finest-grained, having been formed from fine-grained source rocks, e.g., shales and siltstones. This diagram is a fusion of ideas of H. E. Blatt and the writer -- December 1963.

Note: Two geologic realms where straight extinction occurs, at opposite ends of the scale: (1) crystallization from magma or solution, (2) recrystallization of a solid. In between is the realm of undulose quartz. Undulosity is a function of size as well as strain. Thus schists have so much straight quartz because they are the finest-grained, having been formed from fine-grained source rocks, e.g. shales and siltstones. This diagram is a fusion of ideas of H. E. Blatt and the writer--December 1963.

Explanation of page 76. Quartz is more diagnostic of formative conditions rather than specific lithologic types of source rock, but the two are somewhat associated, which led to the original confusion. The following are recognizable quartz types in terrigenous sediments:

A. Special Varieties

1. Volcanic quartz, mainly recognized by shape.

2. Vein quartz, mainly recognized by abundance of vacuoles.

3. Reworked sedimentary quartz, reworked overgrowths.

B. Definite Metamorphic Types

1. Fine polycrystalline quartz, crystals finer than .1-.2 mm, usually with straight extinction because of their small size; straight borders. Origin: recrystallization probably in the solid state. Source: schists, recrystallized siltstones and sandstone, recrystallized chert or sheared vein quartz, etc.

2. Sheared quartz, undulose with crenulate boundaries and two or more elongate individuals. Origin: intense deformation without recrystallization. Source: gneisses, metaquartzites, shear zones in granites, crushed veins, sheared sandstone, etc.

C. Equivocal Types, indicative only of the latest structural event, not indicative of any particular source rock.

1. Single crystal units with straight to slightly undulose extinction (under 2° on U-stage?). Any relatively unstrained or recrystallized source rock: granite, vein, coarse metaquartzite or schist, etc. This is the dominant quartz type in most sandstones, and becomes much more abundant in the finer grain sizes. On prolonged abrasion all other quartz types except this are probably selectively eliminated.

2. Single crystal units with stronger undulose extinction. Any relatively strained rock: granite, vein, coarse metaquartzite, coarse restrained schist, etc.

3. Semi-composite quartz. Any source rock that has been weakly strained and subsequently polygonized; or a comb-structured vein.

4. Coarsely polycrystalline quartz with non-sutured boundaries. Most likely from metaquartzites, but can also come from finer-grained granites, coarser schists, etc.

IV. Authigenic Quartz

These varieties of quartz generally appear to form from solutions at low temperatures and pressures: either in sea water at the time of deposition, from connate water, subsurface water, or ground water. They generally have no connection with magmatic or hydrothermal waters (although such waters can form vein quartz).

Overgrowths form on the quartz grains of sandstones or limestones, especially those of Paleozoic age. They are in optical continuity with the sand grain nucleus, and are separated from it usually by vacuoles, sometimes by clay or hematite; sometimes there is no visible demarcation line and the presence of quartz overgrowths must be inferred by the apparent interlocking of adjoining quartz grains. If pore spaces are large and there is little quartz precipitated, the overgrowths are idiomorphic with nice crystal faces; if too much silica is precipitated so that the pore spaces are completely filled, the overgrowths have irregular boundaries. Overgrowths must be searched for under high power and are often missed because they are difficult to see. Luminescence Petrography reveals them in some examples. Overgrowths formed by direct precipitation usually lack inclusions; those that form by replacement of carbonate always have carbonate inclusions. Zoned quartz overgrowths are common in silcretes and may indicate vadose conditions with alternate wetting and drying.

Idiomorphic crystals averaging .05-.15 mm are common in limestones, where they grow by replacement of the calcite, hence are often loaded with microcrystalline calcite inclusions and have pitted surfaces. They may or may not have sand grain nuclei, and frequently replace fossil or oolite grains.

Nodular Masses of quartz are found occasionally in dolomites. They may form cauliflower-shaped bodies from 0.5 mm to a foot or so in diameter, or even thin beds. They are made up of flamboyant, crudely radiating quartz, and form chiefly by replacement of evaporite nodules. Many have pseudo-cubic zoned quartz (McBride and Folk, 1977, JSP), and associates are length-slow chalcedony. See Milliken (1979 JSP).

Vein and Geode quartz occur as cavity fillings. Crystals are often idiomorphic (mosaic if many are crowded together), and have straight extinction or sometimes develop undulose extinction even without the application of stress.

The time of origin of these sedimentary quartz types has not yet been solved and is a very difficult and economically important problem. Obviously the movement of fluids through a formation will be affected by how tightly that formation is cemented; thus it is essential to determine when cementation occurred as that will impede migration of fluids. Furthermore if one could determine why certain parts of a formation become cemented while others remain porous, one would have a valuable prospecting tool.

These are several possible stages in the history of a sediment wherein cementation might occur. It is not known yet which stages are most important; all stages probably occur in one formation or another. The following classification is based on that of Krynine.

1. Syngenetic Stage

   A. Early Syngenetic (Contemporaneous). Cement develops by precipitation from freely migrating, essentially normal sea water while the sediment is at the sea bottom or at most buried a few feet.

B. Late Syngenetic (Shallow Burial). Cement develops while the sediment is buried anywhere from a few feet to perhaps several hundred feet or more beneath the sea floor. Precipitation occurs from connate (trapped) sea water, which is essentially stationary or at most moved a short distance.

2. Epigenetic Stage

A. Deep Burial. Cement develops while the sediment is buried a thousand feet or much more beneath the sea floor, but prior to any structural deformation of the beds. Precipitation occurs from subsurface water which may have moved long distances through the formation or adjoining formations, and which may be a mixture of fresh water with connate water. Some silica may be supplied by waters pressed out of shales on compaction. Source of silica can be the Si liberated when montmorillonite changes to illite, or from solution of buried opaline skeletons.

B. Post-Structural Deformation. Cement develops during or after uplift or structural deformation of the depositional basin. "Deformation" here includes anything from gentle warping to intense folding and faulting; the essential point is that now sediments are no longer being laid down in the same basin. Source of the silica may be far-travelled subsurface waters; solution of detrital quartz grains under pressure provides a small amount of silica (the so-called Riecke principle, actually first described by H. C. Sorby in 1863; see Sibley and Blatt 1976 JSP).

C. Hydrothermal. Occurs generally during or after structural deformation, from waters heated by magmatic activity and carrying in solution products sweated off from igneous bodies. These may penetrate a formation and be carried along with the migrating subsurface waters.

D. Weathering. Occurs when the sediment is nearing the present ground surface; silica may be carried by ground (meteoric) waters.

For any given formation, deciding when it became cemented with quartz is a very difficult problem. Petrographic work must be supplemented with extensive field work; in fact field work is probably the best method of attack. The microscope must be used however to determine how the cement formed (by replacement or direct precipitation); the nature of diagnostic inclusions which might indicate temperature or composition of the precipitating solutions; whether the cement has spread the sand grains apart by pressure of its growth; the relation of the cement to joints, stains, alteration or other time markers. But field work combined with intelligent sampling is generally required to obtain the areal and stratigraphic distribution of cement, which is generally the key to the problem. For example, if a formation is cemented uniformly over tens of thousands of square miles, the cement is probably syngenetic, formed from sea water at the same time as or just after the sand grains were being deposited. Hydrothermal sources are rather easy to establish or eliminate by noting if the formation is cemented only in the vicinity of known hydrothermal activity; widespread cements cannot be caused by this process. Whether or not the cement has formed on deep burial can be determined by reconstructing the structural history of the area in which the formation occurs; parts of the formation never deeply buried then should lack cement. Cements quite frequently are related to the structural geology of a formation; many are cemented tightly in the vicinity of faults (where shattering has aided migration of silica-precipitating fluids), and some are cemented in tightly folded areas but loose in non-folded ones. In large scale folds a thermal gradient may be established (because of the general increase of temperature with depth), and this affects the solubility of minerals and may cause selective cementation at anticlinal crests or synclinal troughs. Badly jointed areas (as at the crests of folds) may also become tightly cemented because of easier fluid migration. Weathering may be easily determined as the cause by noting correlation of cementation with the actual outcrop of the formation versus its condition in deep cores or spots inaccessible to weathering. If one sets out to solve a problem of this type, sampling is the chief concern. He should take weathered outcrop samples and deep cores; sample folded areas, jointed areas, faulted areas, and flat-lying areas, obtain samples near and far from hydrothermal activity; study the correlation of grain size of the sand with cementation; the relation of cemented zones to overlying and underlying formations, etc.

In general, older formations are more likely to be cemented with quartz than younger ones; yet there are many uncemented Cambrian or Ordovician sands, and some Cretaceous or Tertiary ones are tightly cemented with quartz. In general, deeply buried or strongly folded sands are more apt to be cemented, but many horizontal, unfaulted beds that have never been buried over a few hundred feet are also cemented tightly. The problem is an interesting and difficult one.

Chert and Opal

Sedimentary quartz assumes three forms: (1) megaquartz, a general term for quartz overgrowths, crystals, geode and vein fillings, composed of equant to elongated grains larger than 20 microns; and (2) microquartz, divided into (2A) microcrystalline quartz, forming a pinpoint-birefringent aggregate of equidimensional grains usually ranging from 1-5 microns in diameter (but ranging from a fraction of a micron--in the apparently isotropic cherts--to 20 microns, the arbitrary upper limit); and (2B) chalcedonic quartz, forming sheaf-like bundles of radiating extremely thin fibers, which average about 0.1 mm long but may range from 20 microns to a millimeter long. All three types are transitional in some degree. Both forms of microquartz consist of nothing but finely crystalline quartz containing variable amounts of very minute liquid-filled spherical bubbles, averaging 0.1 micron in diameter, which look brownish in transmitted light and silvery in reflected light (Folk and Weaver). These bubbles are responsible for lowering the index of chalcedony so that it varies from 1.535 in most bubbly specimens up to 1.544 in those with few bubbles. They also decrease the density of microquartz and provide a sponge-like character so that solution is quicker, and this type of quartz consequently weathers faster than megaquartz.

Two length-slow varieties of fibrous silica, quartzine and lutecite, apparently occur as a replacement of evaporites or in sulfate-rich, alkaline environments such as semi-arid paleosoils (Folk & Pittman, 1971, J.S.P.; Milliken, 1979, J.S.P.). Always check optical elongation of fibrous silica to determine this. Zebraic chalcedony (Milliken) is also a common evaporite associate.

Chert is defined as a chemically-precipitated sedimentary rock, essentially monomineralic and composed chiefly of microcrystalline and/or chalcedonic quartz, with subordinate megaquartz and minor amounts of impurities. Common impurities are clay minerals, silt, carbonate, pyrite, and organic matter.

Some cherts contain opal. Chert nodules consist very largely of micro-crystalline quartz, while chalcedonic and megaquartz usually form as cavity-fillings within the chert nodule. In hand specimens, the following varieties of chert are recognized: (1) flint, black due to organic matter inclusions; (2) green jasper, green because of chlorite inclusions; (3) red jasper, red because of hematite inclusions; (4) novaculite, a pure, massive white chert occurring in thick beds, the color due to abundant water inclusions; (5) agate, simply banded chert with different colors, usually a cavity-filling; (6) chalcedony, colorless to pale gray or white, translucent, usually a cavity-filling and usually composed of chalcedonic quartz; (7) moss agate, which is usually chalcedony permeated by manganese oxides in dendritic form.

Chert may form beds hundreds of feet thick; nodules; sporadic replacements of fossils within a carbonate matrix; geode and vein fillings; as a cement in sandstones; and as tiny disseminated specks and dolomolds in carbonate rocks. The latter occurrence is volumetrically far more important than all the others put together.

The origin of bedded cherts is diverse, some are probably carbonate replacements and some are siliceous oozes made of radiolaria or spicules (McBride & Folk, '77-79 JSP). Nearly all nodular cherts result from replacement of carbonate rock while the carbonate ooze was still soft. Evidence for this is the preservation of allochem limestone structures such as fossils, oolites or fecal pellets and bedding laminae in the chert nodule, and disruption of chert by mudcracking and soft-sediment slumping.

Replacement usually results in the formation of microcrystalline quartz, while chalcedonic and megaquartz are usually cavity fillings; however the latter two occasionally form by replacement and the former even more rarely as a cavity filling. A little chert forms by hydrothermal activity and some by weathering.

Opal is isotropic, hydrous silica gel with index varying from 1.40 to 1.47. It has a very strong relief, often thus appears brownish, and sometimes shows weak anisotropism. X-Ray study shows that it contains a little high-cristobalite. It is rare in sediments, and allegedly none has been found in rocks older than the Permian. It may form biogenically as diatom ooze, radiolarians, or spicules; and may form inorganically as cements in sandstones, nodules, weathering crusts, and geode and vein fillings. It is supposed to recrystallize to chert with time. It is often associated with continental volcanic sediments, derived from the decomposition of volcanic ash.

Reworked Detrital Chert

Chert may be eroded out of older limestones and chert beds, and the clastic particles contributed to form a terrigenous constituent of sandstones. Chert may be in turn reworked from these older chert-bearing sandstones. Such reworked chert forms 1-4% of the terrigenous fraction of sediments. It is always diagnostic of an older sedimentary source, hence should be searched for diligently in determining the source area of a given sandstone. It is slightly less durable than quartz (slightly softer, a little more brittle), but despite this, usually occurs in pebbles because the nodules simply started out so big. E. Sneed (U.T.M.A. 1955) has shown that chert pebbles are more difficult to round than quartz because they tend to chip and split; but sand-sized chert rounds faster than quartz because it is softer. It is also somewhat less stable than quartz, because the fine grain size and bubble content presents a large surface area to chemical attack. Weathered cherts develop a whitish outer band or patina due to solution of carbonate or organic inclusions and sometimes spongy solution of the chert itself.

It is possible to set up a classification of chert types for detailed work, based on internal structure (i.e., pellets, oolites, fossils, etc.), grain size of the microcrystalline quartz, color or inclusions. Chert types have been split in great detail by workers in insoluble residues, who find it an invaluable aid in correlation.

Chert, when metamorphosed, grades into fine-grained metaquartzite. Unfortunately, many fine-grained and/or silicified volcanic rocks resemble chert when broken to sand-sized grains, and these must be distinguished carefully. Often the volcanic grains may have a few surviving feldspars with lower index, or may show some trace of phenocrysts.

Feldspar

Feldspar is an important mineral of sediments, and a good clue to the interpretation of paleoclimate. The following feldspars are the ones found in sedimentary rocks (see also J.V. Smith, 3 vol. set, and Van der Plas (1966) on identification).

Orthoclase, $KAlSi_3O_8$, monoclinic. The name adularia is often applied to low-temperature, vein-filling orthoclase but both are identical in all properties. Both may contain some $NaAlSi_3O_8$.

Sanidine, $KAlSi_3O_8$, monoclinic. Differs only from orthoclase in that it has a small 2V and different optical orientation, grains are clear and are usually free of bubble inclusions, in contrast with orthoclase. Sanidine is the high-temperature form, found in lavas. It may also contain considerable $NaAlSi_3O_8$.

Microcline, $KAlSi_3O_8$, triclinic, grid twinning usually with tapering twin lamellae. Often contains considerable $NaAlSi_3O_8$, thus grades into anorthoclase.

Anorthoclase, $NaAlSi_3O_8$ and thus grading into microcline. Triclinic with very close grid twinning, distinguished from microcline by optic orientation; found in soda-rich volcanic rocks.

Plagioclase, a continuous series ranging from Albite ($NaAlSi_3O_8$) through Oligoclase, Andesine, Labradorite, and Bytownite, to Anorthite ($CaAl_2Si_2O_8$). May contain a little $KAlSi_3O_8$. Triclinic with albite twinning, and twin lamellae are straight and parallel. Metamorphic plagioclase is untwinned.

Perthite is a wormy or patchy intergrowth of albite in orthoclase or microcline. It is common in granites and pegmatites.

The feldspars have a hardness of 6, and three cleavages. All the K-feldspars have slight relief with all indices below balsam. Extremely sodic albite (limit An_2) has all indices below balsam; between An_2 and sodic oligoclase (An_{17}) the three indices straddle balsam; and plagioclases more calcic than An_{17} have all indices above balsam. Plagioclases more calcic than An_{30} even have indices higher than quartz. Birefringence of the common feldspars is notably lower than quartz, so that interference colors are gray-white to gray.

	N_x	N_z	Bn	Remarks
Orthoclase	1.517	1.524	.007	2V 70°, (-)
Sanidine	1.517	1.524	.007	2V very small, (-)
Microcline	1.518	1.525	.007	
Anorthoclase	1.523	1.529	.006	
Albite (An_0)	1.525	1.536	.011	Alb. ext. angle 13 to 19°
Oligoclase (An_{20})	1.539	1.547	.008	Oligo, ext. angle 0 to 13°
Andesine (An_{40})	1.550	1.557	.007	And. ext. angle 13 to 23°
Labradorite (An_{60})	1.560	1.567	.007	Labr. ext. angle 28 to 39°
Quartz	1.544	1.553	.009	Uniaxial (+)
Canada Balsam	1.537			

All the feldspars are biaxial, and all have large optic angles except sanidine. The K-feldspars are all optically negative, but plagioclase varies from positive to negative. For details of optic orientation see standard texts, preferably Winchell. Stains are also very useful in identifying feldspars.

Inclusions (other than alteration products--see below) are not all common in feldspar. The scarcity of primary mineral inclusions is really marked. Intergrowth of two feldspars (perthite) and intergrowths with quartz are rather common as detrital grains, however.

Hints on distinguishing feldspar from quartz. Plagioclase and microcline are easily distinguished by their twinning. Orthoclase is much more difficult, however, and is very often missed even by experienced petrographers. In identifying orthoclase, use high power. The diagnostic features are (1) Becke line--orthoclase is distinctly lower than balsam, quartz distinctly higher; (2) because of its lower index, orthoclase looks very pale pinkish or brownish (with fracture lines being bluish), while quartz looks very pale bluish. With practice this becomes the easiest method of identification; (3) Most feldspar grains show cleavage, quartz does not; (4) Feldspar is commonly clouded with bubbles or alteration products, quartz is usually clearer; furthermore, in feldspar the bubbles are elongated or arranged in vague grids or lines parallel with crystal directions, while in quartz they are irregular; (5) Feldspar has distinctly lower birefringence; with practice, this becomes a valuable method of identification; (6) Orthoclase is biaxial-negative with large 2V, quartz is uniaxial positive. Untwinned plagioclase, important in some metamorphic rocks, should be identified by figure. Do not confuse the quartz flash figure with the feldspar figure.

If feldspar and quartz are coated or embedded in hematite, opal, carbonate or clay matrix, those methods based on index of refraction fail, and the easiest method is to dissolve or crush the rock and mount the loose grains in 1.540 oil, then count them. Or one can also cruise around the edge of the rock chip where the sand grains are in contact with Balsam.

Availability. Feldspar forms 5-15% of the terrigenous fraction of sediments. Feldspar is by far the most abundant mineral in igneous rocks, and its much lessened volume in sediments is because of the ease of weathering and lack of durability. Most feldspar comes from granites and gneisses, where orthoclase and microcline are dominant with sodic plagioclase (usually oligoclase) a poor third. Pegmatites contain much more microcline. Volcanic rocks contribute small quantities of plagioclase and a little sanidine. In general, orthoclase is the dominant feldspar in sediments because of its great availability; microcline is a close second despite the fact that it is more stable than orthoclase; sodic plagioclase is a poor third; and calcic plagioclase is almost lacking. Formations in which plagioclase is abundant are consequently unusual. Thus, if a formation contains more plagioclase than K-feldspar, suspect a volcanic source--especially if the plagioclase crystals are zoned. If a formation contains very little or no feldspar, people often assume that it was because of prolonged abrasion or a high rate of weathering; however, it may just as easily mean that little or no feldspar was available in the source area. Schists, phyllites, slates, or older sediments contribute little or none. Occasionally feldspar can be reworked from an older feldspathic sediment into a younger one; but nearly all feldspar comes from a primary igneous or metamorphic source.

Durability. Feldspar is softer than quartz and has a much better cleavage, hence abrasion reduces size and increases roundness much more rapidly. Thus the relation between the size of quartz and feldspar and the roundness of quartz and feldspar is one of the most valuable keys to the history of the sediment and in determining direction of the source area. In youthful sediments quartz and feldspar will be the same size (or feldspar may even be larger since it starts out in the parent rock as larger crystals), and of essentially similar shape. As the sediments suffer more abrasion, feldspar becomes smaller and rounder than quartz; by the time quartz becomes well-rounded, the feldspar is usually much finer. Of course, during the process of abrasion the feldspar continually diminishes in quantity relative to quartz; thus one may start with a good arkose composed of angular grains of quartz and feldspar of equal size, and simply by abrasion alone end up with a supermature orthoquartzite in which all the grains are well-rounded, and the feldspar is reduced to a very small percentage concentrated in the silt size while the sand fraction is nearly all quartz (Folk).

Stability. Feldspars are unstable under deuteric, hydrothermal, and weathering conditions. Four types of alteration are important: kaolinization, sericitization (or illitization), solution, and vacuolization. Kaolin forms as tiny flakes throughout the feldspar grain; the flakes have an index slightly higher than the feldspar, and give a gray birefringence. They do not impart a cloudy appearance to the grain, as the brownish turbidity that textbooks assign to kaolin is actually caused by bubbles. Most kaolin forms on intense weathering by the complete removal of K, but sometimes forms hydrothermally. Sericite and illite form tiny flakes with indices higher than feldspar, giving a white to yellow birefringence and often having crystallo-graphically-controlled distribution. They are most commonly formed hydrothermally but also result on weathering if it is not intense (as in semi-arid climates). Water-filled vacuoles are responsible for the cloudy, brownish appearance of altered feldspars, and vacuolization is the most common type of alteration (Folk). Under high power the vacuoles are brownish (an effect of the index) with index much lower than feldspar. They form on weathering, on hydrothermal alteration, and also are a primary inclusion trapped within the crystallizing feldspar (thus feldspars forming in hydrous environments like granites or pegmatites contain abundant water-filled vacuoles, while the feldspar of volcanic rocks is clear because of the relatively anhydrous environment). T.W. Todd and D.N. Miller, Jr. (U. T. 1955) have discovered post-depositional cavernous alterations of K-feldspars with development of huge holes inside the feldspar grains. Todd (1968 J.S.P.) has studied the complexities of feldspar weathering in soils. Elsie Begle (U.T. 1978) has found much evidence of solution of feldspars in soils with production of deep channelways and skeletal grains.

Feldspars may become altered (1) in the source rock through deuteric or hydrothermal activity; (2) in the soils of the source area by weathering; (3) post-depositionally, either on deep burial by migrating connate waters, or when the sedimentary rocks are exposed to the Recent surface and weathered. Post-depositional alteration may be suspected if weathered feldspar grains tend to occur together, as in the neighborhood of veins, stylolites, or other weathered surfaces; or if well-cemented rocks contain fresh feldspar and porous portions contain weathered feldspar. Comparison of the freshness of feldspar overgrowth with the feldspar sand grains also helps. In some sedimentary rocks, feldspars are partially to completely replaced by calcite; rarely, they are replaced by quartz.

Considering the several kinds of feldspar, microcline is the most stable under weathering conditions, with orthoclase next. Sodic plagioclase is apparently more unstable under surface conditions, for in streams and soils on peneplaned igneous areas there is very little plagioclase, even though it is common in the source rocks. In Llano (Texas) plagioclase dissolves much more rapidly than microcline (Begle). Calcic plagioclase is extremely unstable, thus in sediments usually occurs as the result of volcanism, whereby it has short-circuited any soil-forming process.

Climatic, Tectonic and Physiographic Interpretation based on Feldspar. This is a complex subject, because of the relative durability and chemical stability of quartz and feldspar, together with the confusing effects of rate of weathering and rate of erosion. P.D. Krynine has shown that the main clues are grain roundness, indicating length of time available hence the rate of erosion, the ruggedness of the topography and ultimately the tectonic framework; and average degree of alteration of the feldspar together with homogeneity of alteration of the feldspar. By homogeneity is meant: "do all the feldspars of one species show about the same degree of alteration, or does the same species show a mixture of fresh and weathered grains?" It is normal in sediments to find orthoclase more weathered than microcline or sodic plagioclase, but the thing you must look for is a mixture of fresh plus weathered grains of the same mineral species (Krynine).

A humid climate with rugged topography produces abundant feldspar in angular, coarse grains with a wide variation from fresh to heavily weathered (because the vigorous streams can cut through the intensely weathered mantle to fresh bedrock). If the climate is dry the feldspar will all be pretty fresh, as was observed by Mackie (1899), Judd (1885), and Wade (1911, QJGS). Well-rounded, fresh feldspars if present in moderate to abundant quantitites are excellent indicators of an arid climate and peneplane conditions. These are common in some of our best ancient dune sand blankets, associated with dune cross-bedding and evaporities. The criterion seems to work so well that it may somewhat irreverently be referred to as the "Dogma of the Immaculate Feldspar". If topography is low, and the climate is humid and warm, nearly all the feldspar will be weathered away. These relations are shown in the graph below, in which "Intensity" is to be determined, given the other factors.

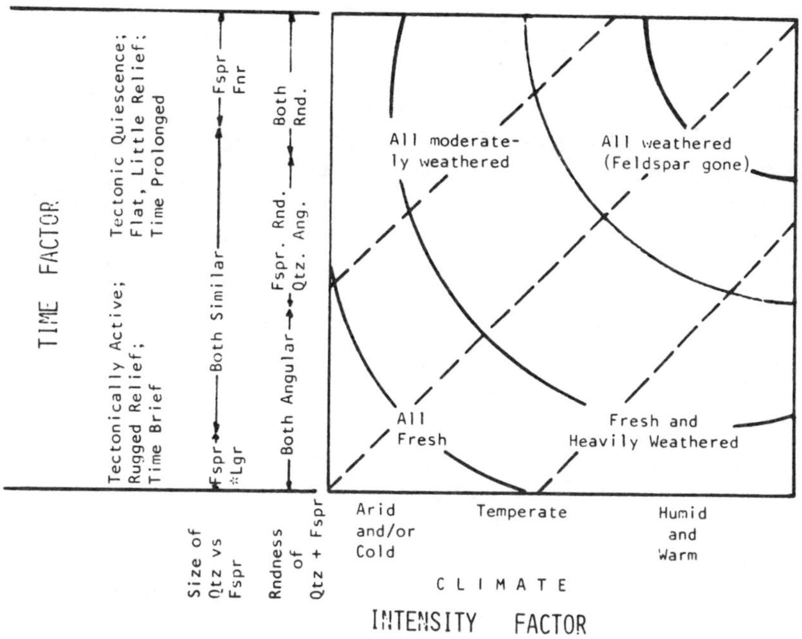

Dashed lines represent contours on homogeneity of weathering, varying from very homogeneous (at the upper left) to very heterogeneous (at lower right); solid, curved lines indicate average amount of weathering (all fresh at lower left, all weathered at upper right) and also indicate roughly the abundance of feldspar, increasing amounts being found toward the lower left.

Authigenic feldspar is common in marine sediments. Orthoclase, microcline, and albite have been formed authigenically, either as overgrowths on feldspar sand grains in clean sandstones, or as overgrowths or new crystals without nuclei in carbonate rocks. In limestones they grow by replacing carbonate (cut across fossils, oolites, etc.) under low temperatures and probably low pressures; in sandstones they form by direct precipitation, again usually under conditions of low temperature and pressure. Authigenic feldspar frequently contains inclusions of carbonate, liquid, or organic matter; sometimes authigenic feldspar of one species may grow over a nucleus of a different feldspar type, e.g., microcline on an oligoclase nucleus. Occasionally the overgrowths are complexly twinned, but usually they are free of twinning.

Occasionally feldspar forms in non-marine, clay-rich rocks (shales or clayey sandstones) by replacement of the clay. Potassic solutions are required, supplied either by connate or hydrothermal solutions. In some tectonic non-marine arkoses, the original red clay matrix has been replaced by feldspar in the form of overgrowths and felted masses of crystals. On burial, some zeolites alter to feldspars especially in volcanic sediments.

Volcanic Rock Fragments. These are a quite diagnostic constituent of sediments, and represent sand-sized bits of the aphanitic groundmass of lava flows hypabyssal intrusives or volcanic ejectments. They are made up usually of a felted mass of tiny lath-like feldspar crystals, occasionally with some ferromagnesian minerals. They are rather soft and highly susceptible to weathering so usually do not last long in sediments. However, "eugeosynclinal" sandstones are full of them.

Volcanic Glass. These particles are formed usually by explosive eruption of siliceous magmas. The glass, if unaltered, is isotropic and has an index well below balsam (more basic glasses have higher indices, and exact measurement of the index can be used to determine the silica content of the lava). Altered glass may be recrystallized to very fine-grained feldspar or microcrystalline silica, or may be changed to clays (usually montmorillonite) or zeolites or may be dissolved and partially re-filled. Glass usually shows characteristic shard-like forms (representing the curved surfaced of the bubble-filled viscous lava) which may be preserved even after total alteration of clays. Hence, Bentonites sometimes show shard ghosts.

Large Micas (> 20 microns)

Muscovite and biotite are hydrous potassium aluminum silicates, with biotite containing ferrous iron and magnesium in addition. Chlorite contains less silica and lacks potassium, and is an alteration product chiefly of hydrothermal or metamorphic origin. Biotite may become leached to a pale golden color through loss of iron on weathering ("leached biotite"), or under hydrothermal attack or weathering may turn from brown to green ("green biotite"), an intermediate stage in the passage of biotite to chlorite probably under reducing conditions. All these species are monoclinic, flaky minerals with biaxial, usually negative figures and small 2V. Sections cut across the flakes are pleochroic with the dark direction N-S, and show parallel extinction, length-slow, with maximum birefringence in 2d order reds, blues and greens. Basal sections are very deceptive inasmuch as they show no cleavage, are almost isotropic, and show no pleochroism, always giving the darkest color.

	Color	N_Y	Br
Muscovite	colorless	1.58	.036
Biotite	brown	1.60-1.66	.040-.060
Leached biotite	pale brown	1.60-1.66	.040-.060
Green biotite	green	1.60-1.66	.040-.060
Chlorite	green	1.57-1.60*	.000-.010**

*Less common forms of chlorite have N_Y up to 1.67.

**Chlorite commonly shows anomalous birefringence--peculiar rich blues, gray-blues, russet browns, or sickly yellowish khaki color.

Mica in sediments rarely contains inclusions of zircon or rutile; usually it is free of inclusions.

Availability. Large micas, although often conspicuous in hand specimens, rarely form more than 2% of the rock even in the most micaceous specimens; altogether, it forms probably less than 1/2% of the terrigenous fraction of sediments. The boundary line between micas and the coarser clay minerals such as illite and sericite is arbitrarily set at 20 microns, so this percentage is subject to considerable revision depending on the size limit adopted. Granites contribute chiefly biotite, with some muscovite in the more acid types. In basic rocks and volcanics biotite is almost the only primary mica present. Pegmatites contain chiefly muscovite, with a little biotite. The greatest source for the micas is, however, in the metamorphic rocks where shales are built up into phyllites and schists consisting of biotite, muscovite and chlorite. Chlorite is also a frequenty hydrothermal mineral, forming by the alteration of ferro-magnesian minerals. Chlorite also forms by low-temperature subaqueous weathering of mafics.

Volcanic biotite is a very important variety. In general, although biotite is probably the most important mica in the source rocks, muscovite is far more abundant in sediments by a ratio of about 4:1. This is a result of the chemical instability of biotite. Hence, when in a sedimentary rock biotite is more abundant than muscovite, two causes may be responsible: either (1) erosion rate far overbalanced the rate of weathering in the source area, or (2) the sediment has received contributions from volcanic rocks or volcanic ash.

Volcanic biotite usually occurs in flakes showing partial hexagonal outlines, and occurs usually in sporadic beds through the stratigraphic section, corresponding to sporadic ash falls. It commonly appears thus in the insoluble residues from limestones.

Durability. Despite its low hardness of 2-3 and the good cleavage, mica is a very durable mineral, extremely difficult to round and almost impossible to fracture. This is becasue it is so elastic; also because of the flaky shape it tends to dodge around grains in transport. Rarely, formations contain rounded flakes of mica but this requires exceptional conditions of prolonged gentle abrasion. Modern dunes of southern California contain coin-shaped rounded micas (Sims).

Stability. Moscovite is stable except under very warm and humid conditions. Biotite is unstable, first becomes bleached through loss of iron, then if under reducing conditions below the water table may lose alkalies and pass into chlorite or vermiculite. Under oxidizing conditions both biotite and chlorite are believed to pass into limonite and clay minerals.

Depositional Characteristics. Because of its flaky shape mica behaves hydraulically as a much smaller particle, therefore large mica flakes are usually washed out of the coarser sands and deposited with fine silts. The percentage of mica varies radically from bed to bed depending on grain size. Flakes are very well oriented parallel to the bedding in most cases. Randomly-oriented micas indicate slumping or some disturbance of bedding.

Authigenic Micas. On metamorphism, clay minerals may be built up into muscovite, chlorite, or biotite. Similar but not so profound changes take place on the sea floor or during deep burial of the sediments. Chlorite is the chief mineral built up in this way (see section on clays), but the flakes are seldom larger than 10 microns. On deep burial, during incipient metamorphism, or in faulted or hydrothermally-soaked zones chlorite commonly forms as tiny flakes, sometimes replacing quartz and feldspar; biotite and sometimes muscovite form when alteration becomes more intense. Biotite and chlorite may also form authigenically as crusts or small flakes in ordinary sediments, but these occurrences are quite rare.

Metamorphic Rock Fragments (MRF's)

Metamorphic rock fragments consist of sand- or gravel-sized pieces of slate, phyllite, schist, metaquartzite, and occasionally gneiss (gneiss is usually too coarse-grained to be recognized as such when broken into sand-size particles). They are the most characteristic and defining components of phyllarenites, formerly called "graywackes." They are composed of fine-grained quartz, micas, and sometimes clays. It is somewhat difficult but necessary to distinguish them from clay pellets or shales; this can be done by the fact that the MRF's are coarser grained, better oriented and harder due to recrystallization; shale fragments tend to be deformed by neighboring grains to a greater extent. Blatt believes it is almost impossible to distinguish shale fragments from slate fragments, and calls them both URF's unless definite evidence of mica recrystallization is visible. For convenience I would lump the whole spectrum under MRF's, even though this does somewhat diminish the "source-area" concept of rock classification. No data has yet been collected, but it is suspected that fragments of slate are more abundant than fragments of shale because of their superior hardness; shale disappears rapidly during transport (Blatt).

Availability. Metamorphic rock fragments form 5-15% of the terrigenous fraction of sediments, and are derived from low-rank to high-rank metamorphic rocks exposed in areas of intense folding.

Durability. Metamorphic rock fragments are quite durable if composed largely of quartz, but quite soft and easily abraded if composed chiefly of fine micas. As abrasion continues on an assemblage of MRF's, the softer ones become rounded rapidly and reduced in size, eventually being broken down into clay which is winnowed away, while the hard metaquartzite fragments persist. Todd (U.T.M.A., 1956) shows in the Carrizo and Newby sands that MRF content may fluctuate from 0 to 30% in successive thin beds, the result of differential abrasion and sorting. Beach sands consequently have much less abundant MRF's than fluvial sands from the same source. Upon very long abrasion, it is possible that metaquartzite may abrade faster than single grains of quartz, and supermature sediments hence may contain relatively less metaquartzite (Blatt). For durability of RF's, see Cameron and Blatt, '71 JSP.

Normally, softer micaceous rock fragments abrade so rapidly that if they are found together with well-rounded quartz it indicates a mixed metamorphic and older sedimentary source, because they could not persist under the prolonged abrasion required to round the quartz.

Since MRF's are usually derived from foliated rocks, the fragments are generally elongated or platy.

Stability. Metamorphic rock fragments are moderately stable chemically because they are made up of quartz and mica, two stable minerals. However, under intense weathering (e.g. Georgia) the MRF's disaggregate into their components, quartz silt and fine micas and then disappear from the sand fraction.

Depositional Characteristics. Metamorphic rock fragments have essentially the same specific gravity and shape as quartz, thus travel with it. On long transport, MRF's wear down faster than quartz, hence finer sands may have more MRF's than coarser coarser sands. Compaction tends to squash some of the softer, more micaceous fragments, reducing porosity significantly.

Sedimentary Rock Fragments (SRF's)

Fragments of older sedimentary rocks are common in conglomerates. Many sands, also, may contain reworked fragments of siltstone, shale, limestone, or fine sandstone. Generally these fragments are not resistant to abrasion, so if present in important amounts indicate brief transport. The abundance of limestone fragments (CRF's) may be used as a climatic indicator. Some sands (e.g., Oakville) are made up largely of limestone grains, as are the Tertiary (molasse) sandstones and conglomerates of the calcareous Alps, Carpathians, etc. Such CRF-rich rocks are known as calclithites.

Clay Minerals

For detailed material on clay minerals, see Grim, Millot, Thorez, Weaver, Velde or other standard works. This is just the merest outline of a few significant facts.

There are three definitions of clay: (1) based simply on size, including anything finer than 4 microns be it true clay minerals, quartz, calcite, pyrite or any other substance; this is the "clay" of the grain-size analysis workers, of the sedimentologists and oceanographers; (2) based on composition, defined as one of the hydrous aluminum silicates belonging to the kaolin, montmorillonite or illite groups and also including fine-grained chlorite and vermiculite; (3) the petrographic definition, which includes under the general term "clay", the true clay minerals listed in (2), plus sericite and fine-grained muscovite, biotite and chlorite if finer than about 20 microns, and even the hydrous aluminum oxides "bauxite" and gibbsite.

Clays are sheet-structure silicates closely allied to the micas; nearly all clays are crystalline (allophane is the only common amorphous one), and all are biaxial negative, with small 2V, length-slow, and if colored are pleochroic with the darker direction N-S. Because they are so similar in these properties, they are difficult to identify microscopically. Study is hindered by their fine grain size (most true clay minerals are finer than 2 microns), by the fact that most naturally-occurring clayey rocks contain intimate mixtures of several different clay minerals, and by the abundance of impurities and stains. Furthermore, the appearance in thin section may be changed by thickness of the slide, by saturation with various materials which may change the refractive index, or by difference in orientation.

Because of the difficulty of microscopic study, clays are commonly identified by X-Ray techniques. The best technique is to get the composition by such instruments, then go back to the thin section armed with this information and study the genetic relationships and associations of the minerals as they occur in the rock (for example, X-Ray might show that the rock contained 20% quartz, 60% illite and 20% chlorite, but it would take a thin section to tell whether (1) the quartz occurred as silt laminae, the chlorite was partially replacing the quartz, and the illite formed the bulk of the shale; or (2) illite in the form of pellets formed the bulk of the rock, chert cement (showing up as quartz on X-Ray) occurred between the pellets, and several chlorite-filled veinlets traversed the rock.

Typical chemical composition of the common clays is shown in the table below, adapted from Grim. It will be seen that in passing from muscovite (sericite) through illite to montmorillonite, the K_2O content drops in discrete steps from 12% to 7% to less than 1%, and that kaolinite and chlorite also lack K_2O; also that illite is high in Fe while most montmorillonite is Mg-rich; and that chlorite is very rich in Mg and Fe and quite low in Si. Water content is omitted in this table.

Some "clays" are very rich in iron. Most "glauconite" is iron-rich illite, "nontronite" is iron-rich montmorillonite, and "chamosite" is iron-rich kaolinite. These three minerals often form as pellets, or even sometimes oolites.

	Gibbsite	Kaolin	Illite	Muscovite*	Montmorillonite	Chlorite
SiO_2	--	44-47	47-52	45	50-56	20-30
Al_2O_3	30-35	37	22-30	38	16-20	15-25
Fe_2O_3FeO	--	0.5-1	2-7	--	1-4	1-40
MgO	--	0.3	1-4	--	3-6	3-35
CaO	--	0.3	0-1	--	1-3	0-3
K_2O	--	0-1.5	6-8	12	0-0.5	--
Na_2O	--	0-0.5	0-1	--	0-2	--

*Sericite is probably just fine-grained slightly impure or K-deficient muscovite.

Although clays are difficult to identify with the microscope, an educated guess can be made with careful work. Chlorite may be recognized by its color; montmorillonite is the only common clay with index below balsam; kaolin, illite and sericite are all above balsam, but kaolin has very low gray birefringence, and the other two appear as bright white to yellowish flakes under crossed nicols; illite and sericite are impossible to differentiate microscopically except that sericite is a little coarser (hard to distinguish by X-ray too, except sericite has sharper peaks). Gibbsite is found only in highly aluminous rocks.

	N_Y	Bn	Color and Remarks
Gibbsite	1.55 - 1.58	.020	Colorless, 20-45° extinction
Kaolin	1.55 - 1.57	.006	Colorless
Illite	1.55 - 1.63	.02 - .05	Colorless to very pale green
Sericite	1.55 - 1.62	.03	Colorless; probably just coarse illite
Montmorillonite	1.49 - 1.52	.02	Colorless (appears spurious brown because of index)
Chlorite	1.57 - 1.60	.00 - .01	Pale to deep green; often anomalous birefringence

It must be realized that the names above refer to groups of minerals, some groups (chlorite, for example) containing a dozen more species differing slightly in composition and optical properties. Properties listed above refer to the most prevalent ranges for the common members of each group. In addition, a great percentage of natural clays are "mixed layer" minerals, composed of intimately interstacked sheets of two or three different species often in no definite proportions--e.g., illite with montmorillonite, or chlorite with montmorillonite. These are actual intergrowths, not simply physical mixtures. An illite-montmorillonite mixed layer clay is easily produced by stripping some of the K+ from a series of illite sheets; this happens readily in weathering. If all the K+ is removed from illite, a material results that gives a montmorillonite X-ray pattern; this is termed "degraded illite" or "weathering montmorillonite." Degraded illites of course have K_2O contents that continuously fill in the range between "good" illite and montmorillonite as shown in the first table.

Origin and Significance. Clays form 25 - 35% of the terrigenous fraction of sedimentary rocks. The clay minerals in a given sedimentary rock may have formed in any one of the following ways; (1) reworked without chemical change through simple disaggregation of older clay-bearing rocks such as shales, clayey limestones, slates, or phyllites; (2) by chemical weathering (soil formation) of minerals containing Al and Si, chiefly feldspar but some from mafic minerals, micas and some from older clay minerals; (3) by subaqueous weathering of volcanic ash; (4) by diagenetic changes taking place on the sea floor after deposition; (5) by diagenetic changes taking place during deep burial, migration of connate solutions, or incipient metamorphism; (6) by intense metamorphism; and (7) by post-diagenetic weathering, taking place after the outcrop is once more exposed to the surface and weathered--thus running the cycle back to (2) again.

(1) Simple disaggregation of older clay-bearing rocks yields chiefly illite, because most shales are composed of illitic clay which is often reworked without much change except for K stripping. Lesser amounts of sericite, chlorite, montmorillonite, and kaolin come from this source. Disaggregation of slates and phyllites yields either illite or sericite depending on the intensity of the metamorphism, with sericite indicating the higher grade; considerable chlorite may also come from this source, but little kaolin and almost no montmorillonite, as these are generally converted into other minerals (illite, sericite or chlorite) on metamorphism. To ascertain whether this is the source responsible for the clay minerals in a given sediment, look for other evidences of a slate or shale source area, such as discrete shale or slate fragments in the associated sands. This is a very important source of clay minerals in Recent sediments and those of Tertiary or younger age, because the continents are now largely blanketed with older clay-bearing sedimentary rocks. An abundance of detrital chlorite and biotite is often associated with volcanism.

(2) Weathering (soil formation) is the ultimate source of most clay. The effect of weathering on clay minerals may be understood if one thing is remembered: the clay minerals because of their tremendous surface area are highly reactive, and will react so as to attain equilibrium with the ions in their environment. Naturally the ions present will depend on two things: (1) what ions are supplied in abundance by the source rock, and (2) are those ions retained in the soil (arid climate) or rapidly removed by leaching (warm, humid climate). Under incomplete leaching K and Mg remain in the soil; if basic rocks are being weathered, abundant Mg ions are liberated and montomorillonite, the Mg clay, forms (possibly sometimes chlorite and vermiculite); if acid rocks are being incompletely leached, abundant K ions are freed and illite, the K clay, results. Under complete leaching, both K and Mg are removed from the soil by percolating ground water, and Kaolin (which lacks both K and Mg) forms. Temperature also affects this reaction, cooler conditions inhibiting formation of kaolin. Under humid tropical conditions and prolonged time, even the Si will be removed from kaolin and gibbsite, purehydrous alumina, is left (but sometimes bauxite can be re-silicated to kaolin). In the humid tropics, basalts give much amorphous clay (allophane). Note that not only feldspars and ferromagnesians break down to clay; even illite, for example, if derived from an older illitic shale and then weathered under warm humid conditions may first break down into "degraded illite" (which is simply the illite structure stripped of most its K ions) and ultimately form kaolin. "Degraded illite" is a rather common product of not-too intense weathering, and gives a montmorillonite-like pattern on the X-ray spectrometer. Weaver has found that this type of "montmorillonite" will take up K and

revert to illite in the lab (also in sea water), whereas most volcanic montmorillonite will not, and this serves to distinguish the two. To determine if weathering is responsible for the particular clay mineral present in a sedimentary rock, the clay mineralogy should be compared with other data, e.g., conditions of feldspars or kinds of fossils as climatic indicators. It has been claimed that a prolonged time of weathering under moderate conditions, may have the same mineralogical results as more extreme heat and moisture for a shorter period. Thus, rarely, kaolin and alumina minerals may form in very old "temperate" soils.

(3) Subaqueous weathering of acid volcanic ash usually produces montmorillonite but sometimes illite; the older bentonites tend to be more illitic, but these may have been montmorillonite to begin with and changed over on diagenesis and deep burial. Volcanism is a major source for montmorillonitic clays, but much montmorillonite develops on weathering (= degraded illite, above). Basic volcanic material often weathers to chlorite. To prove a volcanic origin, look for shards, volcanic fragments, plagioclase, biotite, other mafic minerals, or idiomorphic zircon and apatite.

(4) Marine diagenesis starts altering certain clay minerals as soon as they are dumped by fresh-water streams into the salty ocean, rich in Na, Mg, and K ions. Mixed-layer montmorillonite-illite (i.e., partially to completely degraded illite, the latter also termed "Weathering" montmorillonite) regains its K very quickly from sea water and is partly or completely reconverted to illite. Fixation of K+ to form illite happens very rapidly if the clay is ingested by certain animals and defecated (Jonas). Under certain conditions, it may pick up Mg from sea water and go to chlorite instead. "Volcanic" montmorillonite apparently undergoes no change on deposition in marine waters, or if there is a change it is much slower. Detrital chlorite and illite probably also remain stable in sea water, but minor chemical changes may occur (Ressman & Keller, JSP 1967). The role of kaolinite is uncertain. Because it usually decreases offshore, it was once thought that it altered to other clays (e.g., illite) by reaction with sea water. This view is now largely abandoned. Kaolinite is stable in continental, fresh water deposits and in soils under-going weathering.

Clay minerals in recent sediments are often found to change systematically when traced out to sea; kaolinite being present in continental and nearshore sediments, with illite and montmorillonite farther out. To some, this is the result of marine diagenesis, with the farthest-seaward clays having had more time to react with the saline waters. To others (Riviere; Weaver) this represents a sorting effect, with clay minerals being physically transported for different distances by currents as a result of their differences in particle size or degree of flocculation. The latter school also feels that clay mineral variation both laterally and vertically depends chiefly on sorting processes like that mentioned, or upon changes in petrographic character or climate of the source area; clay minerals, like sand grains, are largely detrital hence chiefly controlled by source area lithology. The other school (Grim) feels that clay minerals are more controlled by the chemical nature of their depositional environment. To determine if depositional environment plays a role in clay mineralogy of a specific formation, determine environment of successive beds by other criteria (fossils, glauconite, sedimentary structures, size and shape criteria) and see if clay mineralogy changes with these different environments--but realize that different clay composition in different environments might be caused by sorting as well as diagenesis. If the clay minerals are constant in beds of different environments, this indicates that the clays are probably controlled by the source area.

(5) and (6) On deep burial and continued, intensified diagenesis, kaolin and montmorillonite appear to be gradually destroyed; if the environment is rich in Fe or

Mg, chlorite develops in their place, and if the environment is rich in K, illite forms. Thus kaolin and montmorillonite are uncommon in pre-Devonian rocks. As true metamorphism begins, both illite and chlorite grow larger, then sericite begins to replace illite in the higher-rank slates, takes over almost completely as illite disappears in the phyllites, and then sericite in turn gives way to muscovite as schists develop. Finally the muscovite abdicates in favor of feldspar and the rock becomes a gneiss or perhaps even a granite. At the same time, if the rock contains much Fe and Mg, chlorite is transformed into biotite in the low-rank schist stage, then may be upgraded to hornblende in the higher-rank schist. In determining if diagenetic processes have happened in a sedimentary rock, look for evidence of recrystallization (i.e., larger-than-normal flakes), or replacement of detrital quartz, feldspar, or calcite by illite or chlorite.

In some cases pore space in permeable sandstones is filled with vermicular hexagonal stacks of water-clear kaolin, encrusting fringes of chlorite or montmorillonite, or large flaky sericite crystals, all probably precipitated from solution just as salt in a beaker in one of the many manifestations of diagenesis. These changes can take place in geologically young sediments that have never been folded or deeply buried, therefore require no extreme heat or pressure. Kaolinite is the most common clay mineral encountered as authigenic pore-fillings in sandstones, but authigenic chlorite is common in marine sandstones with basic volcanic rock fragments.

(7) Post-diagenetic weathering--the same processes occur that operated in (2). The main caution here is to be careful when samples are collected as the clay mineral may be changed on the outcrop of weathering.

To summarize:

Gibbsite is a product of very prolonged humid tropical weathering, as are the other bauxitic minerals.

Kaolin forms on intense weathering where K is removed; a little may be reworked. Possibly diminished by marine diagenesis and destroyed on deep burial or incipient metamorphism. Another school holds that kaolinite does not change on marine diagenesis, and the differences in kaolinite distribution are due to source area changes or to differential sorting.

Illite has less K than muscovite and more K than kaolin, but is a distinct mineral group not gradational to either one. Its structure is less ordered than that of muscovite, and it is responsible for the color of most green shales. It forms in a high-K environment, such as is present in temperate to semi-arid weathering, and develops from other clay minerals on marine diagenesis, deep burial, and slight metamorphism. Much of it is derived from older illitic shales or slates.

Sericite is fine-grained impure muscovite. In sediments the name is virtually abandoned because it is almost impossible to tell from illite. The name is still the one used by hard-rock petrographers to describe the fine, flaky mineral so abundant in phyllites and hydrothermally-altered zones. Most sericite in sediments is derived from erosion of these rocks.

Mixed-layer illite-montmorillonite is a transitional product. It can form when illite undergoes temperate weathering, so that much of the K+ is removed from the structure; and it can also be formed when almost completely degraded illite (stripped of most of its K+) enters sea water and regains much of its original K+ content.

Montmorillonite forms in a Mg-rich environment, most of it by alteration of volcanic ash, some by temperate weathering especially of basic rocks. On incipient metamorphism it may convert to illite or chlorite. Marine diagenesis has little effect on volcanic montmorillonite, except to change adsorbed cations.

Chlorite forms on marine diagenesis, particularly in lagoonal or near-shore marine environments where rivers bearing iron derived from weathering enter the sea. It is also common in marine sands that have received contributions from basic igneous rocks, and forms on diagenesis, deep burial and metamorphism. Much of it is simply detrital, derived from older schists, phyllites, etc. It is possibly formed to a small extent in soils by weathering of micas and montmorillonite.

Talc, pyrophyllite, vermiculite, sepiolite, corrensite, stevensite and other clay minerals occur locally.

Heavy Minerals

Heavy minerals are defined operationally as those with a specific gravity greater than 2.85, the specific gravity of the bromoform liquid used to separate them from lighter quartz, feldspar or calcite. Although over 100 different minerals have been recorded from sediments, they probably form no more than 0.1 - 0.5% of the terrigenous fraction of sediments. Despite their small amount, they are of great value in studying provenance, transportation and weathering history of a sediment and in correlation and paleogeographic studies. They represent the accessory and varietal minerals of igneous and metamorphic rocks, which are much reduced in quantity (except for zircon and tourmaline) as they pass into sediments because they are chemically unstable and also considerably softer than quartz. It is estimated that the average quartzarenite contains 0.05 - 0.2% heavies; phyllarenite 0.2 - 0.8%; and average arkose 1 - 2%, chiefly because the more abrasion or weathering a sediment undergoes, the less unstable minerals it contains.

Common heavy minerals range from 3 to 5 in specific gravity. Because of their heaviness, they usually travel with quartz averaging 0.5 to 1.0ϕ size larger--this difference is known as the "hydraulic ratio" (Rittenhouse) and varies for each mineral species; the value is affected chiefly by the specific gravity but to some extent also by the shape, also by the original size of the mineral grains in the parent rock. Thus if a sand has a median of 2.5ϕ, tourmaline may have a median 2.9ϕ and zircon 3.5ϕ. When a sand such as this is sieved, the heavy minerals fall in the finer size grades, and the heaviest heavies occur at the very finest end; thus when the heavy minerals are mounted one may find that the 4.0 - 4.5ϕ grade has 90% zircon and 10% tourmaline, and the 3.5-4.0ϕ grade has 10% zircon and 90% tourmaline. Sands of different size within the very same bed may consequently have radically different percentages of the several heavy minerals; hence mineral ratios between minerals of different specific gravity or shape are chiefly a function of grain size. One has to use minerals of the same shape and specific gravity--in practice this boils down to making varietal studies of one mineral, such as tourmaline or zircon. Then when comparing ratios between the different varieties one knows that the specific gravity and shape are essentially the same hence there is no hydraulic factor; differences in varietal ratios will then reflect differences in source area lithology.

Shapes of heavy minerals are sensitive indicators to the intensity of abrasion. In comparing beach vs. river sands by grain shape, it is much more efficient to use soft heavy minerals like kyanite or amphiboles rather than hard quartz, as they are affected much more quickly. Angular and rounded tourmaline in the same specimen indicates a multiple source; as would, for example, angular hornblende and round yet harder tourmaline.

Heavy minerals are generally studied in terms of four groups: Opaques, Micas, Ultra-stables, Metastables.

1. Opaque minerals. These generally have very high specific gravity, because of their iron content. Little has been done with them so far. Ore microscopy, in polished sections, may reveal important information, however.

    a. Magnetite and Ilmenite. May form placers of economic value. Very difficult to tell apart except magnetically; both moderately stable, but magnetite may alter to hematite or limonite and ilmenite fairly commonly alters to chalky-looking leucoxene or even fairly large crystals of sphene, anatase, or other titanium minerals. Magnetite often alters to specular hematite, which can not be identified in the microscope; magnetic or X-ray tests must be used.

    b. Pyrite is nearly always authigenic; thus occurs in great amounts in some heavy mineral slides, is absent in most others.

    c. Hematite and Limonite are usually alteration products but sometimes may be detrital. Both dissolve in reducing environments.

    d. Leucoxene is an aggregate of extremely fine-grained sphene, rutile or anatase, and forms as an alteration product usually after ilmenite.

2. Micas. Percentages unreliable because they do not always sink in bromoform. Commonly not counted in heavily mineral studies because of their widely different shape, hence different hydraulic behavior.

3. Ultra-Stable Group. Zircon, tourmaline, and rutile. Because the first two are very hard and inert, (even more so than quartz) they can survive many reworkings. When older sediments are reworked to form younger ones, zircon and tourmaline are about the only ones that can survive. Also, in supermature rocks they are about the only ones that can withstand such prolonged abrasion. Hence they are the backbone of many heavy mineral suites. An abundance of tourmaline and zircon in a heavy suite then means either (1) prolonged abrasion and/or chemical attack has occurred, or (2) the minerals are being reworked from older sediments. The beautiful thing about both tourmaline and zircon is that they are poly-varietal; over 30,000 types of zircon are possible based on color, crystal form, elongation, and inclusions; and tourmaline (Krynine) has been divided into several dozen types each believed to be diagnostic of certain source--e.g. pegmatites, schists, granites, etc. Both tourmaline and zircon are excellent correlation indicators, and idiomorphic zircon is an indicator of volcanism (Callender).

4. Metastable Group.

   a. Olivine is very rare in sediments, occurring only under dry climatic conditions or rapid erosion. Chiefly from basic igneous rocks.

   b. Apatite is moderately stable; commonly it occurs sporadically (abundant in a few specimens, sparse in others), then indicates a volcanic source. Otherwise it can occur in basic to acid plutonic rocks.

   c. Hornblende and pyroxene are moderately unstable; may come from either igneous or metamorphic rocks, but when present in abundance indicate volcanics or metamorphic rocks, such as hornblende schist. Brown Oxyhornblende is diagnostic of a basaltic source; Glaucophane and tremolite are other less common amphiboles, indicating almost certainly a metamorphic source. Pyroxenes are etched and dissolved rapidly by solution after deposition, hence are rare in porous sands.

   d. Garnet may come from plutonic rocks, pegmatites, or metamorphics but in abundance indicates a metamorphic source. Many varieties present, based chiefly on color. Stability variable, depending on the variety; often corroded intrastratally to produce fantastic etch figures. Rapidly dissolved in many porous sands, especially those flushed by fresh water. Garnets bearing chattermark trails are diagnostic of glacial environments (Folk, 1975 Geology).

   e. Epidote, Clinozoisite, and Zoisite indicate a metamorphic or hydrothermal source; they are moderately stable.

   f. Kyanite, sillimanite, andalusite, staurolite are highly diagnostic of a metamorphic source. They are moderately stable, but usually rather soft.

   g. Over 100 others have been reported. For origin and properties see standard works like Krumbein and Pettijohn, or Milner.

<u>Evaluation of Heavy Mineral Content</u>

The heavy mineral content of a sediment is a function of five very complex variables. Usually the aim of the petrographer is to eliminate all variables except one: the lithology of the source area at a given time and place. This he can use as a tool in correlation, and can interpret geologic history by following the source area as it becomes unroofed by erosion and exposes different types of minerals in different amounts. The five variables are:

(1) Lithology of the source area--the ultimate aim of the petrographer. It may be simple (one rock type or assemblage) but more commonly is complex (e.g. sandstone with rounded tourmaline and zircon, overlying schists with angular kyanite).

(2) Differential stability of minerals as they are subjected to weathering in the soils of the source area--to investigate this factor compare the chemical stability of the minerals with their abundance in your sediment.

(3) Durability of the mineral to long-continued abrasion--check hardness against abundance to evaluate this.

(4) Hydraulic factor. Minerals of a certain shape or specific gravity will be carried father away leading to changes in mineral ratios from specimen to specimen. Thus, Brogdon and Bullard have shown that south of the Rio Grande, magnetite decreases southward and hornblende increases southward due to different specific gravity and shape, yet both are coming out of the mouth of the same river; if millions of years later someone drilled a well through these sands, he might think they were coming from a different source area when really it is only the hydraulic factor that is operating. To evaluate this factor, plot mineral specific gravities against mineral abundance in different areas; if hydraulic factors are operating, the points should fall in a fairly straight line or smooth curve, with the platy or elongated minerals systematically displaced to one side with regard to the main trend.

(5) Post-depositional survival factor; on intrastratal solution by migrating connate water, or on surficial weathering some of the less stable minerals (Garnet, Pyroxene, Amphibole, Staurolite) may be destroyed or etched; to check this, compare sealed (cemented) zones with porous zones. In Gulf Coast sediments, unstable minerals appear only in the younger beds, possibly due to intrastratal solution of these minerals in the older beds. Always look for sealed environments to eliminate this factor. Pettijohn thinks this is a very important factor in heavy mineral occurence.

(6) Finally, a statistical error may be mentioned. Erroneous conclusions will nearly always be reached unless standard statistical techniques are used.

All these extraneous variables (2 through 6) can be eliminated if one sticks to varietal counts within the same size grains of the same species; these should have very nearly the same durability, chemical stability and hydraulic behavior and so the only variable operating will be the source area factor.

In studying a heavy mineral suite, one should (1) determine the character of the entire suite, studying shapes, sizes, etc., to determine the general lithology of the source area and study the effect of weathering and abrasion; (2) for detailed work, such as distinguishing different areas of provenance within the same formation or else distinguishing between or correlating two formations, use varietal counts on one or two species, using 5 to 20 varieties per species.

Authigenesis. Magnetite, Tourmaline, Zircon, Rutile, and Sphene can form authigenically in sediments. Tourmaline and Zircon grow as overgrowths on detrital grains, while the others usually form new crystals. Barite, celestite, siderite and pyrite in heavy mineral suites are almost always authigenic.

Carbonate Minerals

There are a great many sedimentary carbonate minerals, but only three or four are important rock-formers--Calcite, $CaCa(CO_3)_2$; Dolomite, $CaMg(CO_3)_2$; Siderite, $FeFe(CO_3)_2$; and Ankerite, $Ca(MgFe)(CO_3)_2$. The Mg and Fe ions are about the same size and substitute for each other readily giving a complete range between (1) dolomite, ankerite, and ferrodolomite, $CaFe(CO_3)_2$; and (2) magnesite, several intermediate Fe-Mg carbonates, and siderite. The Ca ion is larger, however, so that little substitution normally takes place between Ca and Mg--slowly grown inorganic calcite contains at most a percent or two Mg, and slowly grown dolomite contains at most a percent or two excess Ca or Mg over the ideal formula. Organisms, especially those of more primitive types such as crinoids, can build as much as 15 percent $MgCO_3$ into the calcite of their shells without forming any dolomite; more $MgCO_3$ is taken up in warmer sea waters. (Chave). In recent sediments, magnesian calcite and calcian dolomite is not uncommon, because rapid crystallization catches the "wrong" ions. To a lesser extent, Mn is also easily substituted for Mg or Fe, giving rhodochrosite ($MnCo_3$) and kutnahorite (Mn-dolomite).

All these minerals are uniaxial negative with extreme birefringence, rhombic cleavage and hardness 3-4. Calcite and dolomite have one index well below balsam, the other well above so that they "twinkle" on rotating the stage (except for near basal sections, which always give the higher index); siderite has both indices well above balsam, therefore does not twinkle. Furthermore, both ankerite and siderite often are brownish due to iron staining and partial alteration to limonite pseudomorphs.

Calcite probably forms over half by volume of the carbonates, occurring as directly-precipitated microcrystalline ooze (forming irregularly rounded grains 1-4 microns in diameter); as mosaic, xenomorphic crystals which are clear and often twinned, and as fibrous coatings and crusts. In some fresh water environments (caliche, streams) calcite forms tiny euhedral rhombs (Folk, 1974). It forms "lithographic" ooze, cement in sandstones and limestones, oolites, fecal pellets, fossils, vein fillings, and occasionally is a replacement mineral. In some sandstones, it occurs only as scattered patches while the rest of the rock is noncalcitic. These patches usually consist of one large crystal engulfing numerous sand grains, and the rocks are known as "lustre-mottled" sandstones or sand crystals. Aragonite has the same composition as calcite but is orthorhombic with slightly higher indices. It occurs in fossil shells and recent oolites and carbonate oozes; it inverts to calcite eventually, but some Pennsylvania shells are still aragonitic, while some Pleistocene aragonite has already changed to calcite.

Dolomite may occur as a directly-precipitated (?) ooze in crystals 2-20 microns in diameter; as coarser, idiomorphic to xenomorphic crystals replacing limestone; and occasionally as veinfillings or cement in sandstones. It may be distinguished from calcite by the fact that it nearly always shows some evidence of rhombic crystal form, even in grains as small as a few microns; also, it is almost never twinned. When replacing limestone, the allochem structure is sometimes preserved, especially if the dolomite crystals are coarse; but just as commonly it is obliterated. Even in one thin section there may be areas where this structure is nicely shown, and adjoining areas where it is destroyed. "Limpid" dolomite (perfect, exceptionally clear crystals) probably indicate fresh-water mixing, and "Baroque" dolomite (coarse, white, twisted crystals--Pedone 1978) probably indicates S or SO_4 association and higher temperature.

Siderite and ankerite may, like dolomite, form as a replacement, as a directly-precipitated ooze, or as concretions. Ankerite may form beds, reworked pellets or intraclasts in some swampy, deltaic sediments (Rizvi).

Miscellaneous Chemical Minerals

Magnetite is usually a detrital heavy mineral, but it may form authigenically under mildly reducing conditions, especially when associated with glauconite. Under most reducing conditions, magnetite is dissolved. Hematite may be dull red and earthy, or else form blood-red translucent scales; it is the pigment of most redbeds. Oxidizing conditions are necessary both for its formation and for its survival. May form in soils under humid, tropical conditions, or may be precipitated from solution in the sea. In most redbeds, hematite acts only as a stain (i.e., very finely divided, uniformly distributed small particles) and forms a coating on the clay minerals comprising only a few percent of the rock; in other specimens, however, solid masses of pure hematite form as pore-filling cement or replacement of carbonate. Limonite is frequent as a weathering product of other iron minerals. It requires conditions less hot and humid than hematite in order to form in the soil, thus occurs chiefly in temperate soils. Aging converts limonite to hematite, also. In many recent or continental sediments it forms a solid cement, probably as the result of weathering. Walker (1967) thinks red sandstones may develop under semi-arid conditions by alteration of heavy minerals, intrastratally. Mature desert sands are red (Folk, 1976); with aging the red darkens to dull maroon.

Pyrite forms under reducing conditions where sulfur is present, and is commonly associated with organic or petroliferous matter, fossils, etc. It is a very "strong" mineral and can replace anything, even quartz. It may form isolated tiny crystals or huge concretions, the latter usually around an organic nucleus. Melnikovite or hydrotroilite is a very finely-divided black iron sulfide responsible for much of the dark color of recent marine muds.

Glauconite, KAlSiOH with considerable Fe and Mg, varies from an Fe-rich type of illite to an Fe-rich type of montmorillonite. In thin section it is usually some shade of green with a microcrystalline or scaly birefringence. It is supposed to form under marine, mildly reducing to mildly oxidizing conditons on the continental shelf, and is abundant in modern offshore muds. Much of it occurs as fecal pellets; rarely it forms as a pore-filling cement, or replaces calcite. It is considered very diagnostic of marine beds, and the type of glauconite varies with environment, geologic age, etc. (Burst). Collophane ($CaPO_4$) occurs as bones, teeth, scales, shells, conodonts etc.; as turds; and as inorganic replacements of limestone beds, as nodules and oolites.

Evaporites sometimes occur as beds, sometimes as cement in sandstones or as replacement crystals in limestone. Gypsum has low birefringence with index below balsam, and is usually fibrous; anhydrite has indices above balsam, with high birefringence and often forms good crystals; and halite is isotropic, with very low index. Barite and celestite may occur as cement in sandstones or as replacement in limestones; index is above balsam, gray birefringence. Evaporites are frequently replaced by length-slow chalcedony or quartz. Zeolites occur as common cements in volcanic sediments, and may also result from burial metamorphism.

As mentioned previously, zircon, tourmaline, rutile, sphene, anatase and other odd minerals may form authigenically. These often form good stratigraphic markers.

Organic matter, though not a mineral, deserves mention. It occurs as black or deep brownish material, and represents the most resistant humic and bituminous materials such as waxes, lignins, etc., left after decay. It is most abundant where formation is rapid (flourishing life), where burial is fast, where bacteria are not active (bacteria eat it up and destroy it)--stagnant, reducing waters are regarded as most favorable. Modern marine muds run 1-2% organic matter; there is very little in sands because organic matter is light and fluffy and travels with the clay minerals. Liquid hydrocarbons--paraffins, naphthenic and aromatic--are now forming in recent marine muds.

PETROLOGY OF SANDSTONES

According to Krynine (Jour. Geol. 1948) sedimentary rocks have three basic properties: mineral composition, texture, and structure. Mineralogy and texture have already been discussed from a piecemeal aspect; now it remains to see how they are integrated in the classification and used in interpreting the genesis of sandstones. Structure is not covered in this syllabus, many excellent texts now exist [e.g. Pettijohn/Potter, (1963-1964)].

Textural Maturity. The texture of sediments has two aspects: (1) description of properties, i.e., determination of the grain-size parameters and grain-size name, measurement of grain shape and description of surface features, and (2) integration of these properties into an assumed sequential development, the four stages of textural maturity. (Folk, J. Sed. Pet. 1951, 1956). This concept proposes that, as sediments suffer a greater input of mechanical energy thorugh the abrasive and sorting action of waves or currents, they pass sequentially through the following four stages:

I. Immature stage. Sediment contains over 5 percent terrigenous clay matrix; sand grains usually poorly sorted and angular.

II. Submature stage. Sediment contains under 5 percent clay, but sand grains are still poorly sorted (σ over 0.5ϕ) and are not well rounded.

III. Mature stage. Sediment contains little or no clay, and sand grains are well sorted (σ under 0.5ϕ), but still not rounded.

IV. Supermature stage. Sediment contains no clay, sand grains are well sorted and well rounded (Waddell roundness over .35; ρ over 3.0). This determination should be made, if possible, on quartz grains of medium and fine sand size.

Textural maturity is thus one of the most important keys to the physical nature of the environment of deposition, since it provides a descriptive scale that indicates the effectiveness of the environment in winnowing, sorting, and abrading the detritus furnished to it. Thus immature sediments accumulate in loci such as flood plains, alluvial fans, or neritic or lagoonal environments where current action is either weak or deposition is very rapid so that sediments do not have a chance to be subjected to input of any mechanical energy after deposition. Supermature sediments, on the other hand, indicate deposition in loci of intense abrasion and sorting, such as beaches or desert dunes, where energy is constantly being expended on the grains.

Because it is supposed to be an index of original sedimentational conditions, one ignores later effects of diagenesis--i.e., discounts authigenic clay minerals that have crystallized in formerly open pore spaces, ignores chemical cements, and considers shapes of grains underneath overgrowths if any. To determine textural maturity, follow this key:

1. Determine the percentage of clay (including micaceous hash finer than coarse-silt size, .03mm).

    a. If the rock has over 5 percent clay, it is IMMATURE.

    b. If the rock contains less than 5 percent clay, go to (2).

In general, immature rocks have very little original visible pore space, and little chemical cement; submature (and higher) rocks have the pores either open or filled with chemical cements like quartz or calcite.

2. Determine the sorting of the rock.

   a. If it is not well sorted with σ over 0.5ϕ (mm diameter ratio over 2.0), it is SUBMATURE.

   b. If it is well sorted with σ under 0.5ϕ, go to (3).

The sorting borderline between submature and mature is depicted in the bottom figure on the next page, which represents a set of spheres with a phi standard deviation of exactly 0.5. Compare your sample with the figure to see if it is well enough sorted to be mature.

3. Determine the roundness of quartz grains of sand size.

   a. If the grains are subangular to very angular on Powers' scale (ρ under 3.0, it is MATURE.

   b. If the grains are subround to well rounded (ρ over 3.0) it is SUPERMATURE.

The figure below represents a sand with roundness of 3.0, the borderline between mature and supermature.

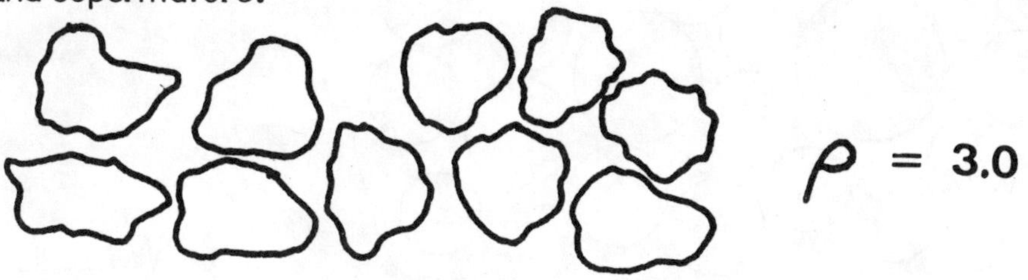

TEXTURAL MATURITY FLOW SHEET

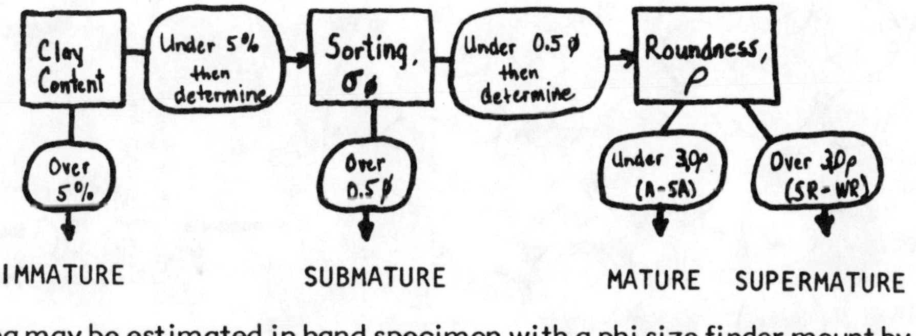

Sorting may be estimated in hand specimen with a phi size finder mount by determining what phi diameter has one-sixth of the grains by area larger than itself; then determining the phi diameter that has one-sixth of the grains (by area) smaller than itself. In this way, one has estimated the 16th and 84th percentiles of the distribution. Subtract these two diameters and divide the result by 2 to get the standard deviation. For example, if the one-sixth limits are 1.2 and 2.8ϕ, σ is 1.6/2 or 0.8ϕ. In thin section, the same process of mentally determining the diameters at the 16th and 84th

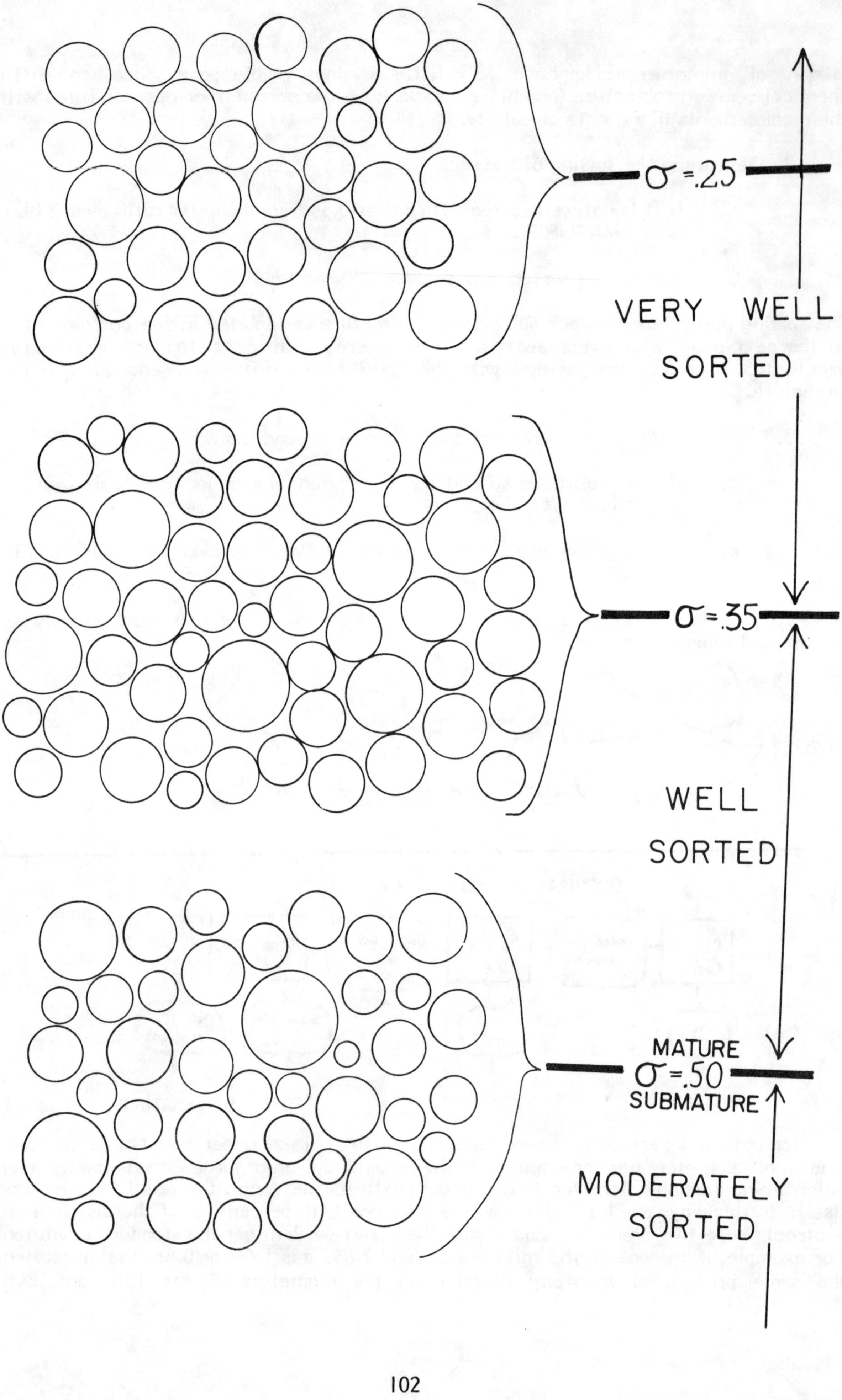

σ=.25
VERY WELL
SORTED
σ=.35
WELL
SORTED
MATURE
σ=.50
SUBMATURE
MODERATELY
SORTED

percentiles is gone through, except the grains are measured in millimeters. The millimeter values can be converted to phi values by using the chart on page 23 and treated as above; or more roughly one can divide the two millimeter values and use the ratios in the following table (for example if the one-sixth limits are .40 mm and .15 mm, the diameter ratio is 2.7, and the rock is moderately sorted and submature).

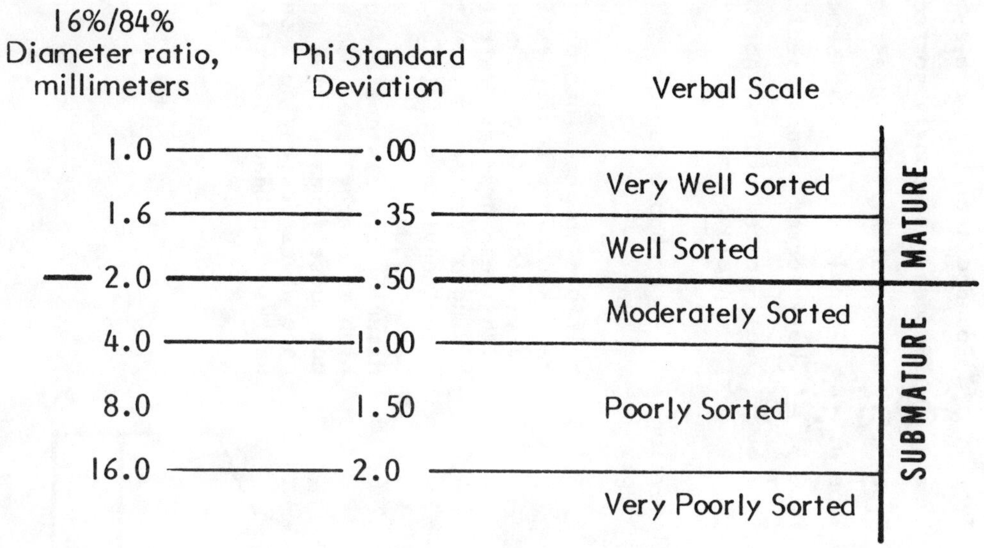

Unfortunately, it is a common tendency to overestimate the standard deviation values of well-sorted sands. Hence these values can be more conveniently obtained by comparison with the set of standard deviation comparison images, previous page.

In determining maturity, there is one exception to the numerical limit $\sigma = .5\phi$ advocated above. Some sands that would be considered poorly sorted as whole, consist of two distinct modes which are in themselves well sorted. These examples should be called "bimodal mature" or "bimodal supermature" (depending on the rounding), considering them to be genetically well sorted for the purpose of maturity terminology, although numerically they would not be. Many of these are desert deflation products (Folk, 1968 XXIII Int. Geol. Congress, Praha Cedskoslovensko).

Textural Inversions occur when well-sorted or well-rounded grains occur in a clay matrix, or when a sediment is composed of poorly sorted but well-rounded grains. These are very valuable in interpretation because they indicate mixing of the products of two energy levels. Sorted or rounded sand grains in a clay matrix often occur in lagoons behind barrier bars, where the sand grains are blown off the beaches or dunes (where they achieved their sorting and rounding) and mixed with the lagoonal clays by storm winds or waves. Final deposition occurs in a low energy environment, therefore texturally inverted sediments are classified as to the lowest stage of maturity present, which normally represents the latest environment. Textural inversions also occur when older sandstones are eroded to produce a new sediment, for example poorly sorted river sands made up of rounded grains from outcrops of much older sandstone. Some textural inversions may be caused by burrowing organisms; for example, pelecypods or worms could burrow through a nicely interlayered series of well-sorted sands and interbedded clean clays, and make the whole thing into a homogeneous mass of clayey, immature sand. But the presence of these immature sands would indicate that the final environment was one of low energy, or else the currents would have re-sorted the material after burrowing.

TEXTURAL MATURITY

Energy Applied by Waves and Currents

Maturing processes are effective only within black range. At too low energy, they fail to operate; at too high an energy level they are self-destructive (lined range), and thus become responsible for textural inversions.

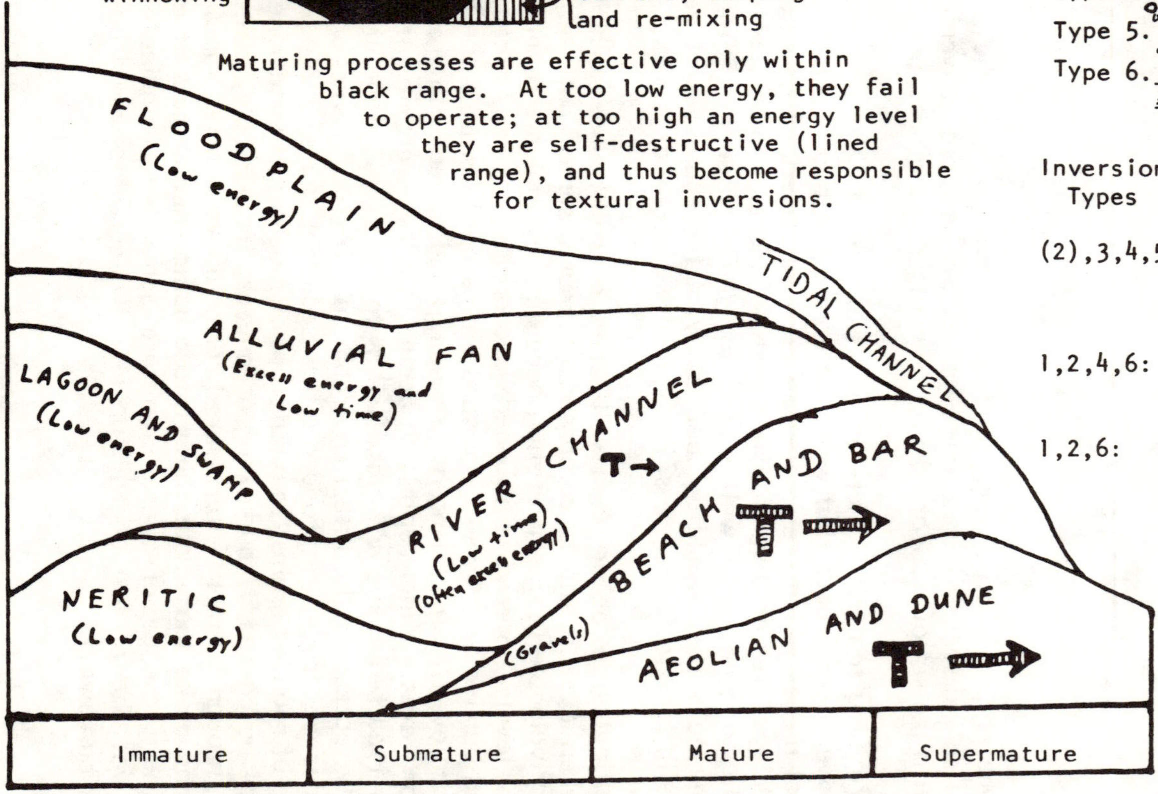

TEXTURAL INVERSIONS

Type 1. Rounded grains in clayey matrix
Type 2. Rounded but poorly sorted grains
Type 3. Abnormal size/roundness relations
Type 4. Bimodal roundness in same size grade
Type 5. Well-sorted bimodal sediments
Type 6. Well-sorted grains (not well rounded) in clayey matrix

Inversion Types	Probable cause
(2),3,4,5:	Multiple source; incorporation of reworked older sediments into primary material (e.g. rounded Є sands with fresh granitic detritus)
1,2,4,6:	Mixing of sediments from two different environments (e.g. barrier bar + lagoon)
1,2,6:	Maturing process operating at too high an energy level hence becoming self-destructive (e.g. hurricane waves destroying a series of well-sorted beach deposits)

Indicates that sediments in the environment in which the symbol appears move to the right (become more mature) if prolonged time is available. The other environments show little change with time. Reasons for the lack of maturity of sediments in some environments are given in parentheses. In a general way tectonic stability correlates with high maturity, instability with low maturity, thus tectonism and environmental influences both react on maturity.

ENVIRONMENTS

NOTE: All above factors affected by geologic history. For example the sediments we find now in the neritic zone may have actually formed when sea level was much lower--ancient beaches are often buried under several hundred feet of water and their characters will be those of beach sands although they are now deep in the neritic zone. Beaches contain gravel if (1) gravel is the result of past geologic history (e.g. glacial); (2) hard rocks outcrop at the coastline; (3) competent rivers carry gravel to the sea.

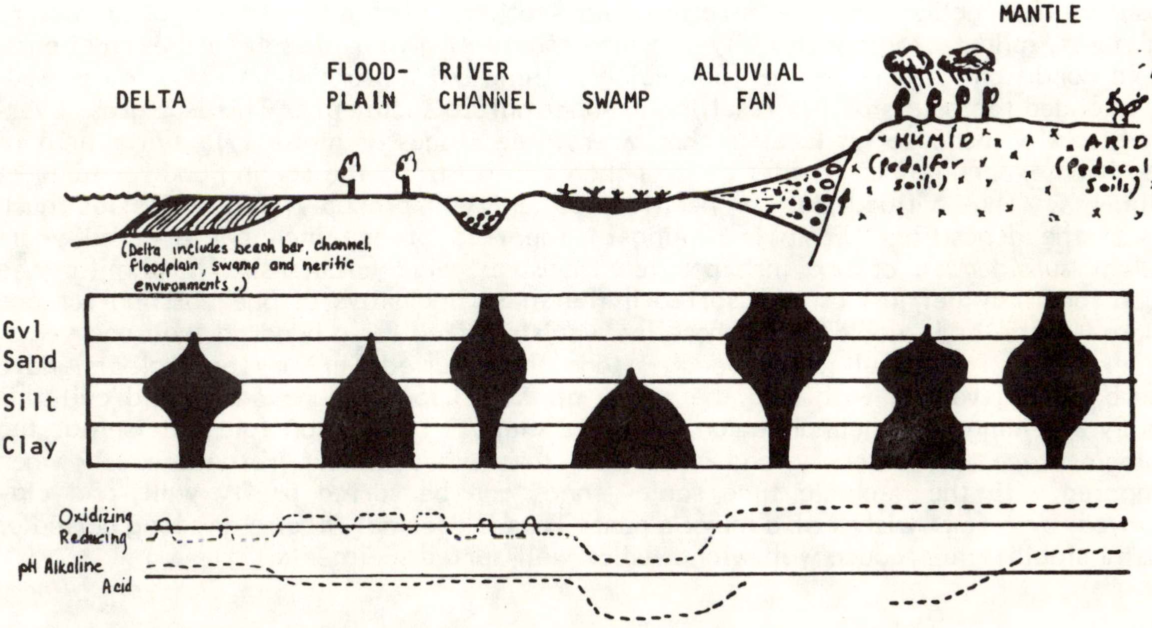

NOTE: Proportion of gravel in the various loci depends on (1) nearness to hard rock outcrops or older reworkable gravel deposits, (2) competency of currents.

The range of maturity encountered in natural environments is estimated in the following figure, thickness of the lenses corresponding to the most common stages of maturity present in each environment. The next figure attempts to show typical grain size for the various sedimentary environments.

It is important to realize that the maturity rating of an environment depends on how much mechanical energy is exerted on a sediment after it has been moved essentially to its final resting place, by currents and waves at the final site of deposition; it does not depend on the energy expenditure required to move it from the source area to the site of final deposition. A flood or turbidity current, for example, may expend a tremendous amount of energy while it is transporting sediment, but once the sediment is dropped it is simply buried by more sediment and never suffers any further sorting or winnowing; thus such sediments have low maturity. Waves on a beach, on the contrary, sort and rework the sediment continually as the tide goes in and out, or as storms and seasonal changes chew up beach sediments and shift them repeatedly.

Krynine has shown that each sediment-modifying process operates at a certain optimum energy level, and this philosophy is certainly true in concept of textural maturity. Thus the amount of energy expended in modifying sediments must be within certain limits; if it is too little, sorting and rounding do not operate efficiently; if too much energy is exerted, the maturity may be destroyed (such as when a sudden storm destroys a well-sorted beach and mixes it with lagoonal clays, or when a mountain cloudburst produces torrents that fracture boulders that had become rounded under a more gentle stream regimen). Thus sudden excess bursts of energy are often responsible for textural inversion. It is a curious thing that the best rounded sands are usually not too well sorted; this may mean that rounding operates best in environments with energy levels that are too high for optimum sorting, where currents, winds or waves are too vigorous and repeatedly mix up and "unsort" previously sorted layers of sands.

Tectonism and Textural Maturity. It has been pointed out above that textural maturity is very largely the result of the environment of deposition (beach, lagoon, floodplain, delta, alluvial fan, etc.) Some workers, however, claim that textural maturity is a direct function of tectonism: intense tectonic activity with rapidly subsiding geosynclines yielding stratigraphic sections composed entirely of immature sediments, mild tectonic instability (unstable shelves) giving all submature sediments, stable conditions (stable shelves) producing nothing but mature sediments, and periods of prolonged tectonic stability resulting in supermature sediments. This is a gross over-simplification because (at least in the lower three stages of maturity) environment of deposition exerts a much greater control than tectonism on the sorting and rounding of sediments. Thus a flood-plain or neritic sediment will probably contain just as much clay if the depositional basin is sinking at the rate of one inch per hundred years (violent subsidence), or one inch per ten thousand years (stable shelf). Similarly, a beach sand may get just as well sorted if the shore line stays at one position for one hundred years, as it would if the shoreline was stabilized for a hundred thousand years. For example, Texas Gulf coast beaches (and their buried Tertiary equivalents) have superb sorting values, with σ in the range of 0.25-0.35ϕ; yet no one would call this rapidly subsiding geosynclinal region a "stable shelf." The reason for that is that the geologic processes of sorting and winnowing take place almost instantaneously when compared with the geologic time scale--sands can be sorted pretty well, and clay removed, by a few swishes of a miner's pan! Thus it does not require any long period of crustal stability to produce well-winnowed or well-sorted sediments.

The role of tectonic stability in producing supermature sediments is less clear. Certainly those times in earth's history that were characterized by very stable and quiescent conditions (upper Cambrian, Ordovician, and Cretaceous for example) have produced abundant supermature sediments; and relatively stable areas of the earth's crust have produced supermature sediments throughout much of geologic time (Canadian and Australian shield sediments, for example). But contrary evidence shows that supermature sands can be produced in one cycle, and in very brief geologic time if the environment is potent enough. This happens in the Ellenburger of West Texas, where beds of supermature sand alternate repeatedly with angular sands of the same grain size, all coming from the same primary granitic-gneissic source area. The identical thing happens in the Silurian of West Virginia, where quartz sands from the same metamorphic and vein source may be very well rounded in one bed and subangular in a bed two feet higher. If periods of tectonic stability are called upon to explain these sequences, then those "stable" periods must have lasted only a small fraction of a gologic period. In the opinion of this writer, then, these rapid alternations are caused by shifts in environment (beach or dune vs. fluvial or neritic, for example) and do not require any prolonged stability of the crust. They probably would not form in a period of such rapid shoreline shifts as we have had in the Pleistocene or Recent (since very little rounding of sand-size grains is going on in present U.S. ocean beaches), but might be formed during non-glacial times in places where the environment was right. The beach-dune environment would have to be relatively stationary at one place (or else sweep through the same spot repeatedly) in order to produce high rounding, but the point is that it need not be stabilized for more than a fraction of a geologic period.

Although the environment of deposition is apparently the immediate controlling factor in textural maturity, the tectonic framework exercises an indirect control by determining which environments shall be volumetrically dominant and which environments shall be rare in a given region or stratigraphic section. As Krynine has shown, the degree and type of tectonic activity does determine a certain preferred (but not necessary) association of source area lithology, relief, geomorphic processes, and rate of subsidence of the depositional basin. These factors in turn integrate to produce preferred associations of environments, because the rate of influx of detritus into a basin, combined with the rate of subsidence of that basin, determines the quantitative distribution of the environments, e.g., proportion of continental to marine facies, or relative importance of deltaic vs. each sediments. And these environments in turn control the textural maturity. For example, in a tectonically active geosynclinal area sediment deposition is rapid with buildup of extensive floodplains, aggrading river channels, large deltas, and thick masses of neritic sediments, possibly in part deposited by turbidity flows and submarine mudslides. In such a tectonic framework, immature sediments will dominate because of the prevalence of low-maturity environments in the region. But local beaches in this complex, although volumetrically minor, may contain extremely well-sorted sediments of high maturity, and some of the river channel and shallow neritic sediments may be pretty well sorted. When intense block-faulting occurs in the continental interior (e.g., Basin and Range province), very rapid erosion and presence of abrupt scarps leads to the production of immense thicknesses of continental sediments, huge alluvial fans, mudflows, bahadas, bolson fill, and fluvial sediments. As a whole then, these deposits will be of low maturity, but occasional river channel sands or lake beaches (e.g., Lake Bonneville) may be quite well sorted. In a tectonically stable shield area, the shoreline transgresses and regresses over large areas because of the flat topography, while subsidence is very slow and little sediment influx occurs. Thus most of the sediments are beach (or dune) deposits, and fluvial sands have only a temporary existence because they are destroyed and reworked by the next marine transgression. The beach sands are packed one against the other by shoreline shifts, to produce a wide sheet of highly matured sand. But neritic sediments (and occasionally-preserved fluvial sediments) may be just as clayey and poorly sorted as the fluvial or neritic sediments produced in the other more active tectonic frameworks.

Thus grandiose tectonic conclusions should not be drawn from one or a few samples, unless it can be shown that they are typical of a large stratigraphic interval. A single bed of very well-sorted sand may represent a beach, but one cannot jump immediately and conclude that the beach sand formed in a tectonically stable framework; it may be just a local beach in a great mass of deltaic sediments formed in a violently subsiding geosyncline. The decision on what type of tectonism is represented has to be made by examining the quantitative distribution of environments over a large stratigraphic interval, and integrating this with other rock features such as mineralogy, thickness distributions, and gross stratigraphy.

Thus, the textural maturity of sandstones is dependent on environment, but volumetric importance of specific environments is determined by tectonic activity. The mineral composition of sandstones is controlled by source area lithology, and this in turn is also affected by tectonism (Krynine). Consequently both properties interact, although they show a considerable degree of independent variation. Their interaction produces what may be called a "main sequence" of sandstone types: most immature rocks, usually formed during periods of crustal unrest, are rich in unstable constituents like metamorphic rock fragments, micas, and feldspars, hence are usually arkoses or litharenites. Most supermature rocks, formed during periods of crustal stability, have suffered enough abrasion or weathering to remove the unstable constituents, and the only thing that remains is quartz; thus they are usually quartzarenites. Sublitharenites and subarkoses frequently represent transitional stages of maturity and mineralogy. There are many exceptions to the main sequence, however, which are very important to recognize as they represent unusual conditions of climate or depositional environment. Mature litharenite, supermature sublitharenites, immature and submature orthoquartzites, and mature and supermature arkoses are quite common examples falling off the main sequence. Any mineral composition can occur in any stage of textural maturity, but certain conditions are preferred in nature.

Krynine's Theory of the Tectonic Control of Sandstone Properties

M. S. Shvetsov, in the 1920's, was apparently the first one to realize that pure quartz sandstones are characteristic of stable shelf areas and quiescent continental shields, or kratons; and that sandstones in orogenic areas are much more complex mineralogically--either arkoses (rich in feldspar), graywackes (rich in rock fragments) or tuffites (rich in volcanics). But Krynine went further and about 1940 set up his system, a three-stage scheme for linking the tectonic development of a continent with deposition of specific mineralogical sandstone types. These were, in chronological order; (I) Quiescence or Peneplanation Stage, resulting in "quartzite"; (II) Moderate Deformation or Geosynclinal Stage, resulting in "graywacke"--i.e., a sandstone rich in metamorphic rock fragments, micas and micaceous clay matrix; and (III) Violent Deformation or Post-geosynclinal Phase, resulting in arkose. For references in which these ideas were presented, see PDK memorial in 1966 JSP. The most accessible references are JG 1948 and AGU Trans. 1951. The following is an ultra-brief condensation of Krynine's ideas.

A typical segment of the earth's crust is assumed to consist of three layers: sediments on top, metamorphics and veins next, and plutonic igneous rocks at greatest depth. Krynine's idea is that increasing amounts of tectonic activity are required to bring these successively deeper layers to the surface so that they can act as source areas.

FOUR-DIMENSIONAL TECTONOLITHOLOGIC CONTINUUM

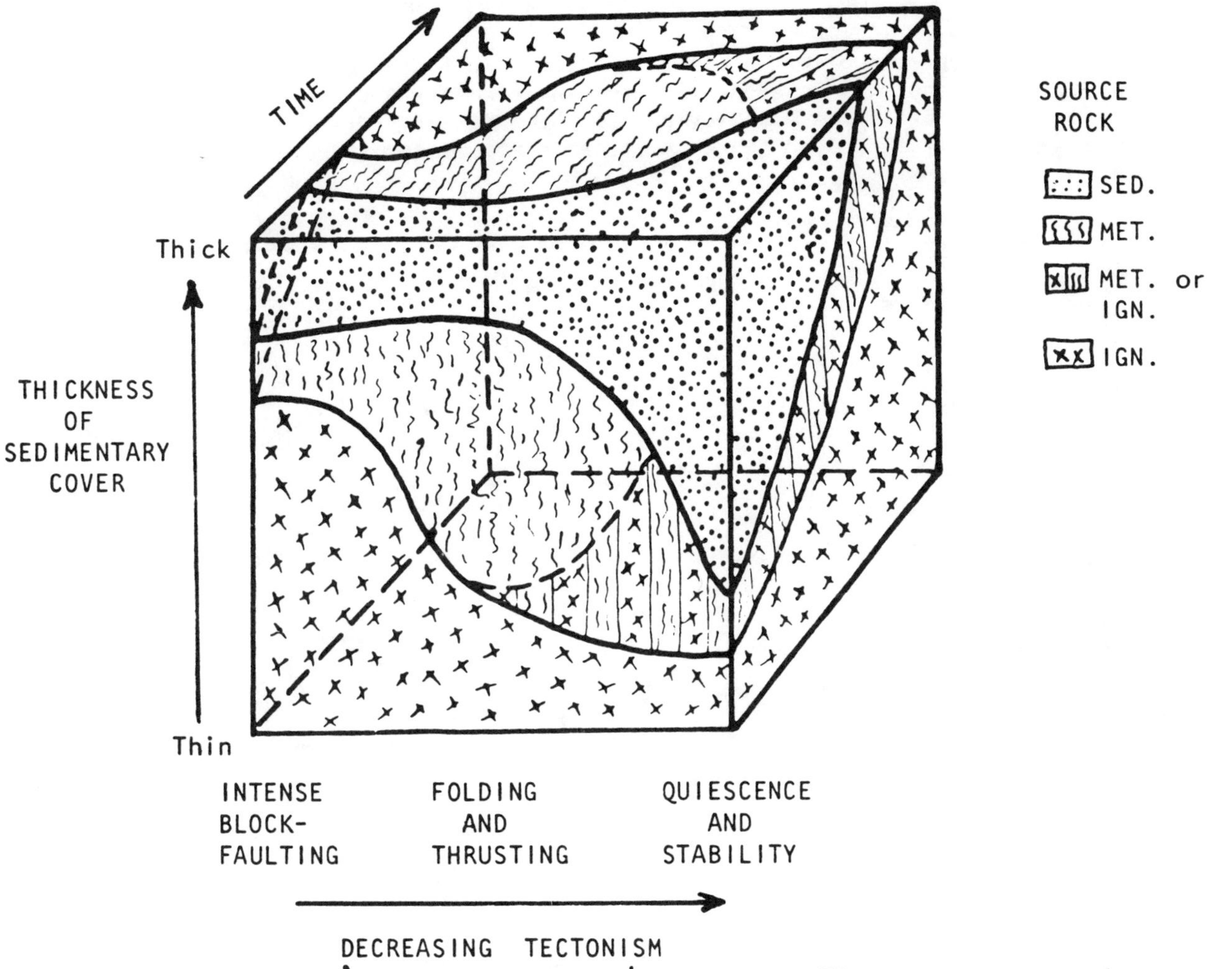

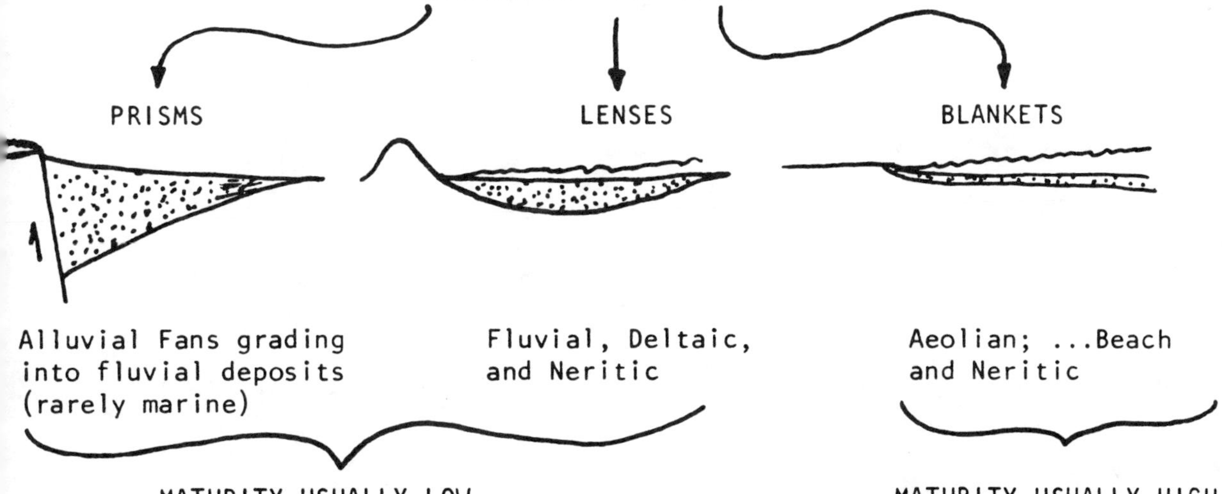

MATURITY USUALLY LOW, increases away from source as size decreases away from source.

MATURITY USUALLY HIGH, uniform maturity and size over wide areas.

PREFERRED ASSOCIATIONS OF THE TECTONIC STAGES

(<u>There are numerous exceptions</u>)

MINERALOGIC CLAN, TEXTURAL MATURITY, AND TECTONIC BACKGROUND

Almost any rock type can occur in almost any tectonic framework, depending mainly on source, but strongly affected by climate, environment, etc. However, there seem to be certain preferred associations as shown below. This is a reasonable estimate, based very largely on North American examples unfortunately. (1963)

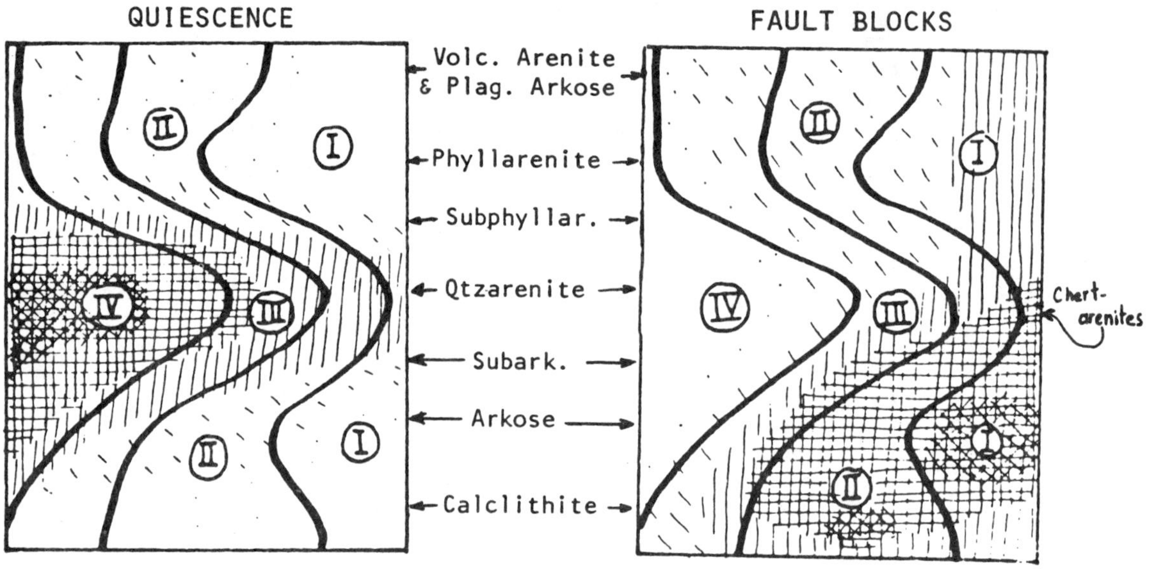

Textural Maturity

I Immature; (clayey) II Submature; (poor sorting) III Mature; (sorted) IV Supermature (rounded)

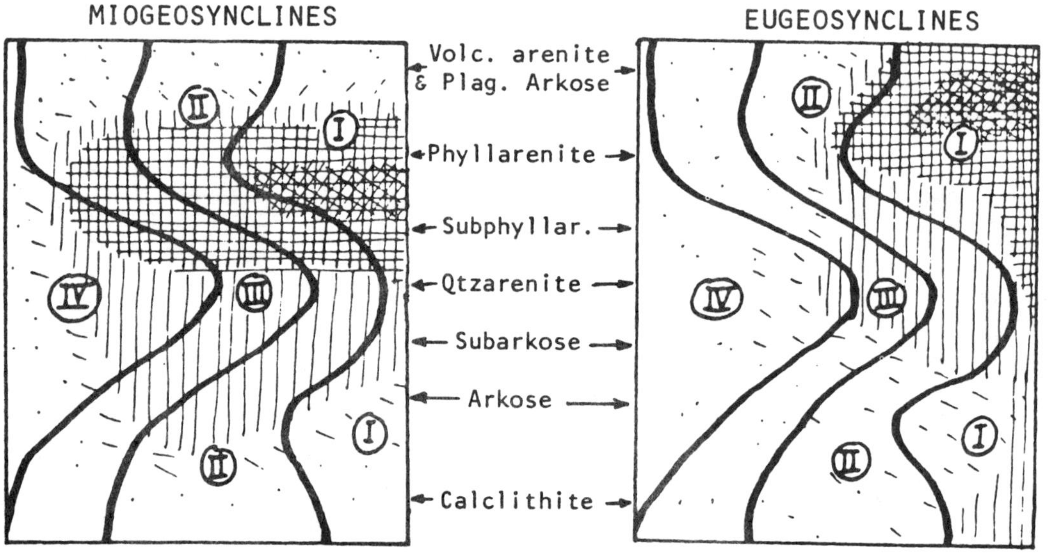

Each background diagram is the same, and intends to show the relative abundance of different maturity classes in each rock clan. Density of shading shows, for a given tectonic background, the estimated abundance of each clan/maturity combination. All rocks within the lined area are common; most densely shaded area shows the "classical" occurrences.

Under quiescent tectonic conditions the top layer of the crust (sedimentary layer) furnishes the detritus. Furthermore, under quiescent conditions, intensely matured beach and dune sediments dominate the stratigraphic section. For both these reasons, then, the resultant sediments will consist largely of quartz (with or without chert), i.e., they will be quartzarenites. If quiescence is prolonged, erosion may eat down to the granitic layer, and some feldspar may enter the sediment.

Under orthogeosynclinal deformation, tectonic activity buckles the metamorphic layer up to provide a source area. Intense folding, thrusting, and dominant horizontal tectonic movements actually create low-rank metamorphic rocks like slates and phyllites out of the folded sediments. Because of high relief and rapid erosion, the dominant environments are those where rapid sediment dumping takes place, i.e., fluvial planes, deltas and neritic shelves. Thus sands are rich in metamorphic rock fragments and are not well sorted; hence, they will be graywackes (now, 1966, termed "Phyllarenites"). Fold mountains (Alps, Pyrenees, Slovak Carpathians) contain abundant phyllarenites.

If violent tectonic movements of a dominantly vertical nature occur, as in block faulting following a period of folding, the deepest layer of the crust, granite and gneiss, is raised to the surface to provide a source area. Environments again are those of rapid deposition, i.e. fluvial plains and alluvial fans. Sediments consequently are rich in feldspar, therefore are arkoses.

Krynine's theory is vastly oversimplified and idealistic, nevertheless it works rather well in many areas and is an excellent guide to stratigraphic thinking. Its big imperfection--and no model is ever perfect--is that it ignores the important part played by the past history of a region, which is a large "chance" variable thrown into the ideal tectonic scheme. For example, consider an area that was in some past age a geosyncline filled largely with shale and that later became orogenized and converted to low-rank metamorphics. If such an area is block-faulted, the pile of metamorphics would be so immense that MRF-sandstones (graywackes of PDK) would be deposited instead of, as theory proposes, arkoses.

Related to this is the absence of any place in the scheme for sedimentary deposits resulting from simple rejuvenation (without differential deformation) of an old source (of whatever lithology). Another weak point is the very minor role assigned to volcanic sediments.

In view of these relatively minor shortcomings to what is basically a sound organized scheme, the following "New Gospel" (conceived in fall, 1965) is propounded as a modification of Krynine's scheme and to take into account these difficulties.

A Concinnity of Depositional Regions

To understand the relation between tectonics and sandstone types, one must first consider gross continental architecture. The following scheme is a development of the work of Stille, Knopf, Kay, Krynine, Krumbein-Sloss, and others. For the record, terms originated by myself are shown thus (*).

In its early stages, a continent may consist of a stable kraton (typically mostly granite or gneiss or high-grade metamorphics), surrounded in part by subsiding regions of sedimentation (geosynclines) with outlying welts serving as most active sources of detritus; most detritus moves centripetally, from the outlying welts towards the kraton.

[Greek spelling is Kraton (= established, stable); introduced in geology to Kober in 1921 as "kratogene," later popularized by Stille as "Kraton."] Thanks to A.M.C. Segör for etymology.

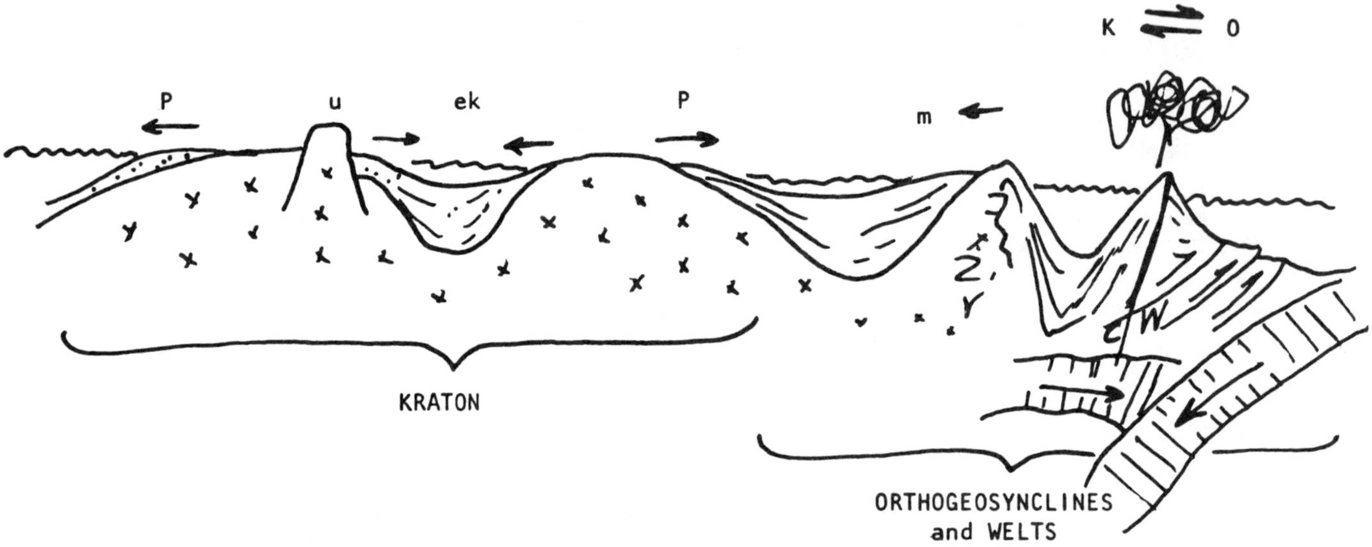

The highly mobile Orthogeosynclinal Belt (H. Stille) contains:

w: tectonically active welt, often volcanic; islands or large lands.

e: Eugeosynclinal belt (Stille), typically deeply subsiding, very thick sediment infill (often by turbidites and so-called "flysch"), rich in volcanics and deep-sea sediments. The outer belt is usually beyond the ken of geologists. Recent work has shown that it is often a trench overlying a subduction zone between two tectonic plates (K⇌O). Between the trench and island arc is the "arc-trench gap."

r: a medial ridge or island chain that may or may not be present as a separation between eu- and miogeosynclinal belts; if it is absent, the two types of orthogeosynclines may be smoothly transitional. The whole depression then is often considered a "marginal basin" and may be produced by "back-arc spreading."

m: Miogeosynclinal belt (Stille), typically less subsidence, thinner, with local basins and highs, typically shallow-marine, paralic to continental; essentially non-volcanic. A strongly uplifted point source may give the exogeosyncline of Kay (e.g., Catskill "delta," New York - Penna.).

Either the mio- or the eugeosynclinal belts may be missing, but if both are present the eugeosyncline is on the oceanward side.

The relatively stable Kraton Region contains:

p: Perikratonic shelves and basins, occurring on the perimeter of the kraton and fed by the kraton; generally thin blankets or basins that are not elongated parallel with kraton margins; shallow-water marine sediments dominant (stable shelves of Krumbein & Sloss).

ek: Epikratonic* shelves, basins and geosynclines (parageosynclines of Stille), developed on the interior of the kraton, away from its margins (auto- and zeugogeosynclines of Kay). Rift Valleys (KΛK) may represent aulacogens. Wider basins result from "Saucer Tectonics."

u: Uplifts may or may not be present adjoining the epikratonic basins.

All of the above features are completely gradational and overlapping, but represent recognizable end-points like the colors of a spectrum.

Later in the history of the continent, the mobile belt is thrusted, folded, squashed, intruded, uplifted, and accreted onto the continent as a relatively rigidified outer frame, which thererafter does not normally undergo further severe horizontal deformation. Thus the original kraton (hedreocraton of Kay) is enlarged to become a new kraton, or neokraton.* Any mobile belts added since Precambian are considered as neokratons. Further geosynclines, basins, sheets, and troughs may then develop atop of and flanking the neokraton, as classified below:

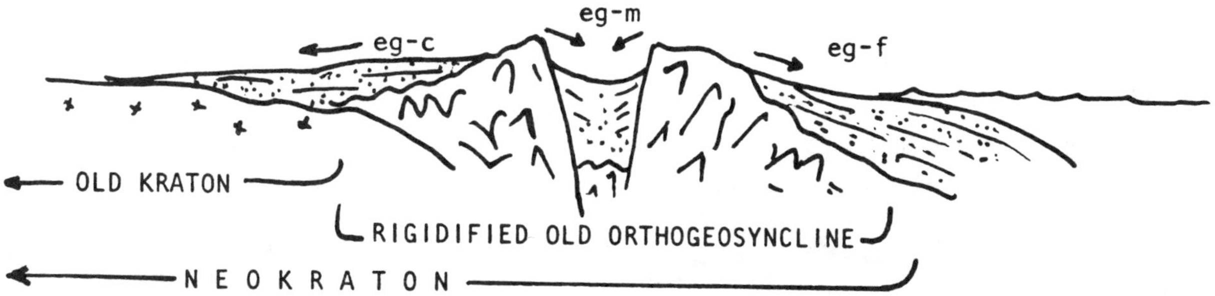

eg: Epigeosynclinal* sheets, basins, troughs, and geosynclines; can be divided as follows:

eg-f: flanking linear regions of thick deposition (paraliageosynclines of Kay--more exactly, "geomonoclines" like Atlantic and Gulf Coast Tertiary, U.S.A.); or more equant basins or sheets (e.g., Sydney Basin, Australia). These often occur on the passive edges of continents, where continent and ocean basin are on the same tectonic plate. (KOΛ).

eg-m: medial regions (including local intermontane basins, fault-block basins or taphrogeosynclines of Kay, and deeply subsiding troughs or epieugeosynclines of Kay; e.g., Newark series, or Nevada graben-fills, etc.) These may be the result of tensional movement between plates, or the initial rifting when a continental plate starts to split. (KΛK).

eg-c: centripetal regions, sediment moving generally from the deformed belt back onto the stable kraton (e.g., Great Plains Tertiary, USA; Cretaceous and later, western Queensland, Australia). Dilution from the old kratonic area common.

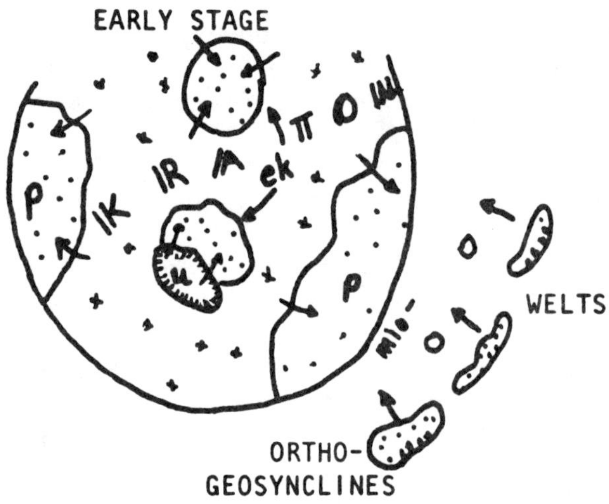

ek - epikratonic basins and geosynclines
u - uplift
p - perikratonic basins, sheets, and troughs
→ transport directions

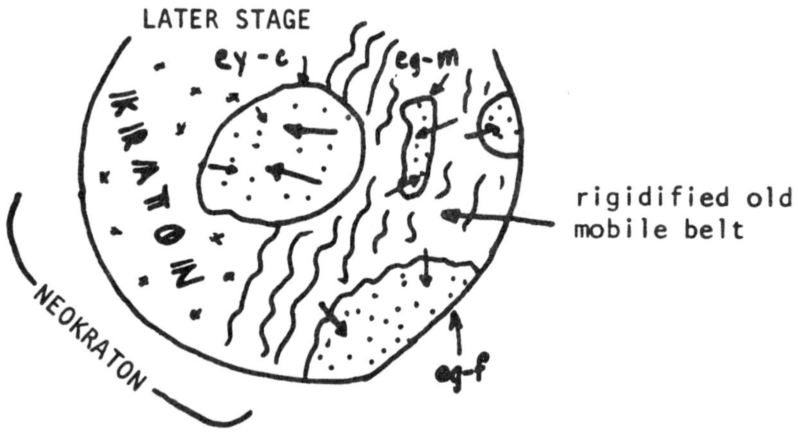

eg - epigeosynclinal basins, troughs, sheets, etc., mostly lying atop the neokraton
→ transport directions

The New Global Tectonics Integrated with Classical Sedimentary Tectonics

The concept of mobile oceanic crust, colliding and splitting continents and plate tectonics has wreaked great changes in ideas of geosynclines and classical "sedimentary tectonics" as championed by the workers of the pre-1960 period. However, one should not be defeatist and say "let's throw out all this garbage on nomenclature of geosynclines," or "tectonics has no relation to sandstone petrology." The two schemes can be satisfactorily integrated; true it is not simple but the steak should not be thrown out with the grease. (See Dott, SEPM Sp. Publ. 19 and 1978 E. Sci. Rev.; and Dickinson, SEPM Sp. Publ. 22).

Surely we recognize that there are deep troughs wherein enormous sections of sediments accumulate, and other areas that are stable with only thin undeformed layers of sediment. And among the sediment-filled troughs there are many different settings; non-volcanic vs. volcanic, occurring in different positions with respect to continents, plates, etc.

What we see in a thin section of a sandstone usually is the endpoint of an incredibly long chain of events extending back into the dimmest geologic past. As we trace causes back things get more and more hazy. Nevertheless it is worthwhile to try to trace this chain of events as far back as we can, to as ultimate a cause as possible. Beginning at the prime source, (1) convection currents in the mantle presumably cause (2) movements of crustal plates. These movements produce (3) zones of surface rocks having different types of tectonic activity--compression, tension, volcanism, etc. These in turn (4) become different types of source terranes: metamorphics, granites, andesites, etc. These in turn (5) produce different types of sandstones as modified by weathering processes, relief, rate of erosion, and differential abrasion or weathering during transport and deposition in various environments. For example, plate convergence produces horizontal deformation (Dh), e.g., thrusts and folds, and this often produces metamorphic rocks, which under rapid erosion, ineffective weathering and lack of abrasion will produce fluvial sediments that are phyllarenites, or beach sediments that will be quartzarenites with a lot of metamorphic quartz. As another example, plate separation produces vertical deformation, i.e., tension and block faulting and either uplift of plutonic basement (Dv) or granitic intrusion, thus common granitic sources, hence alluvial fans and arkoses. These are ideal examples, providing the "main sequence," or the "guiding model." Mother nature is not that simple and we have to realize that variations exist because of random interfering causes, e.g., past geologic history.

Reducing plate tectonics to its basic framework, one can establish the following classes of plate behavior.

I. STABLE PLATE (symbolically, K for continent). Basins of sedimentation atop an unbroken continental plate. Plate character may range from a shield of prolonged neutral or mildly positive nature (Canadian shield), to a mildly depressed perikratonic margin with widespread thin sheets of platform sediment (mid-Continent U.S.); it can have local gentle basins (Michigan) or block-faulted troughs (late Paleozoic, Colorado). Occurring in this framework are perikratonic shelves, and the parageosynclines of Stille (audo-, zeugo-, and taphrogeosynclines of Kay). Sediments may also spill off the edge of a stable continental plate, producing a geocline indistinguishable from KOΛ (see below). Tectonic activity mainly Q_k or Q, (giving carbonates, supermature quartzarenites orsubarkoses) locally R or even rarely D_v (arkoses). Continental plate is not ruptured, hence is strong enough to resist later crumpling at that site, thus no D_h; basins contain only mildly deformed sediments.

IIA. PLATES SEPARATING, <u>Ensialic</u> <u>rift</u> (KΛK, read K split K). A continental plate is beginning to split apart, often because a pair of oppositely-directed mantle currents well up underneath it. However, simatic basement is not yet reached. But a linear geosyncline forms upon the interior of a continent (miogeosyncline). This may be filled by sediments swept in from the stable Kraton on either side (Q_k); often thick troughs of quartzarenite result. Continental plates, weakened along this join, may osculate (repeated kissing-type activity): compressional spasms causing D_h (intense thrusting, folding and even metamorphism with production of phyllarenite clastic wedges), while tensional spasms cause D_v (block-faulting and arkoses). These two phases often alternate and may be interspersed with restful phases, forming quartz-arenites or carbonates. Minor volcanism and intrusion may occur. The Amadeus Basin of central Australia appears to be a good example of incipient plate separation with later compression, collapse and healing. Continental plates may also slide laterally, K/K (read K slide K), producing D_v uplifts and basins. See Lowell 1972 GSA. Aulacogenes are incipient KΛK.

IIB. PLATES SEPARATING, Ensimatic rift $K\Lambda_s K$, read K split-s K). Further cracking apart of continental plates exposes simitic basement. A very important, deep linear "eugeosyncline" is formed; usually vast outpourings of basic to intermediate volcanics, often with deep-water sediments (shales, cherts) and turbidites; sandstones usually phyllarenites, volcanic- or plagioclase-arenites. Some material (often quartzose) may spill in from the stable Kraton (Qk) during phases of relaxation, especially in the "miogeosynclinal" belts that flank the medial "eugeosyncline." Alternating spasms of compression (causing D_h and normally phyllarenites) and tension (causing D_v) are the rule, with abundant volcanism (X). Examples are the Appalachian geosyncline during much of the Paleozoic when it had a kraton-ward miogeosynclinal belt and medial eugeosynclinal belt between the osculating American and European plates. The Ural geosyncline is also probably an example of this.

IIC. PLATES SEPARATED, Oceanic rift produced (KOΛ). Plates have finally separated, osculating ceases, and a substantial oceanic zone lies between the two fragments. Continents continue to drift apart and the ocean widens, the mid-ocean ridge is active, and mantle currents remain in their positions. Sediments are provided by the kraton and the now-rigidified neokratonic frame and tectonized source lands that had been active highs during phase II (KΛK). Detritus spills off the continental margin into the newly-formed oceanic basin, forming a "geocline," or Atlantide geosyncline. Tectonics mostly R or Q, some D_v, little or no D_h; volcanism not generally important. Example: Atlantic coast of North America during Mesozoic and later time. Sandstones not distinctive; dirty quartzarenites common, and some rejuvenation phyllarenites.

IIIA. PLATES COLLIDING, Ocean/Continent ($K \rightleftharpoons 0$, read K crash 0). Basic oceanic crust is thrust under the edge of the continent in a trench and subduction zone. Trench sediments, turbidites, volcanics horribly mingled; intrusion of granites, ophiolites, etc. Common volcanic island arcs, (classic "eugeosyncline") with a kraton-ward miogeosynclinal belt. Pacific type geosynclines of Crook. Mostly continuous intense D_hX (thrusting and folding with volcanism; phyllarenites and volcanic- or plagioclase-arenites), but with occasional resting or even tensional phases. "Miogeosyncline" can be supplied mainly from the kraton side; "eugeosyncline" from the volcanic arc. If volcanics break through on the continent (not a separated island arc), a huge terrestrial volcanic pile may result instead of a geosyncline (e.g., W. Mexico).

Dickinson has proposed that plates may shear (slip sideways) instead of collide, but sediment types are probably not much different; there may be less volcanism. This shear situation may be symbolized K/0 (read K slide 0).

IIIB. PLATES COLLIDING, Continent/Continent ($K \rightleftharpoons K$). The two continental plates were once separated by a substantial ocean. Conditions appear to be not much different than IIIA (K 0) except less volcanism; usually a tectonic land is lifted up at the line of collision. The Himalayas are an example of $K \rightleftharpoons K$. Intense D_h caused by

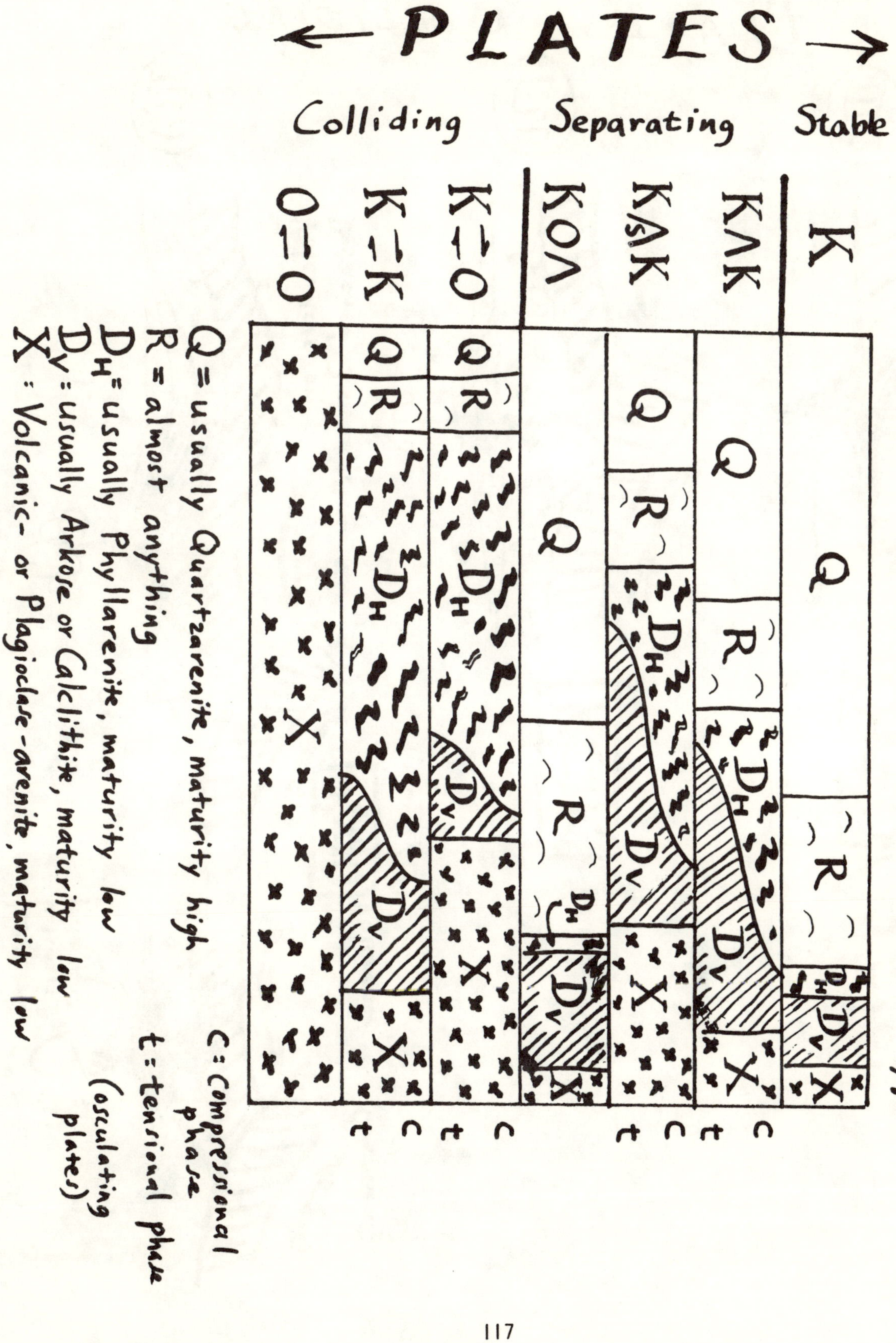

← PLATES →
Colliding
Separating
Stable
Plate Behavior linked to Tectonic Type
Q = usually Quartzarenite, maturity high
R = almost anything
D_H = usually Phyllarenite, maturity low
D_V = usually Arkose or Calclithite, maturity low
X : Volcanic- or Plagioclase-arenite, maturity low
Sandstone Types
c = Compressional phase
t = tensional phase (osculating plates)

PLATES COLLIDING

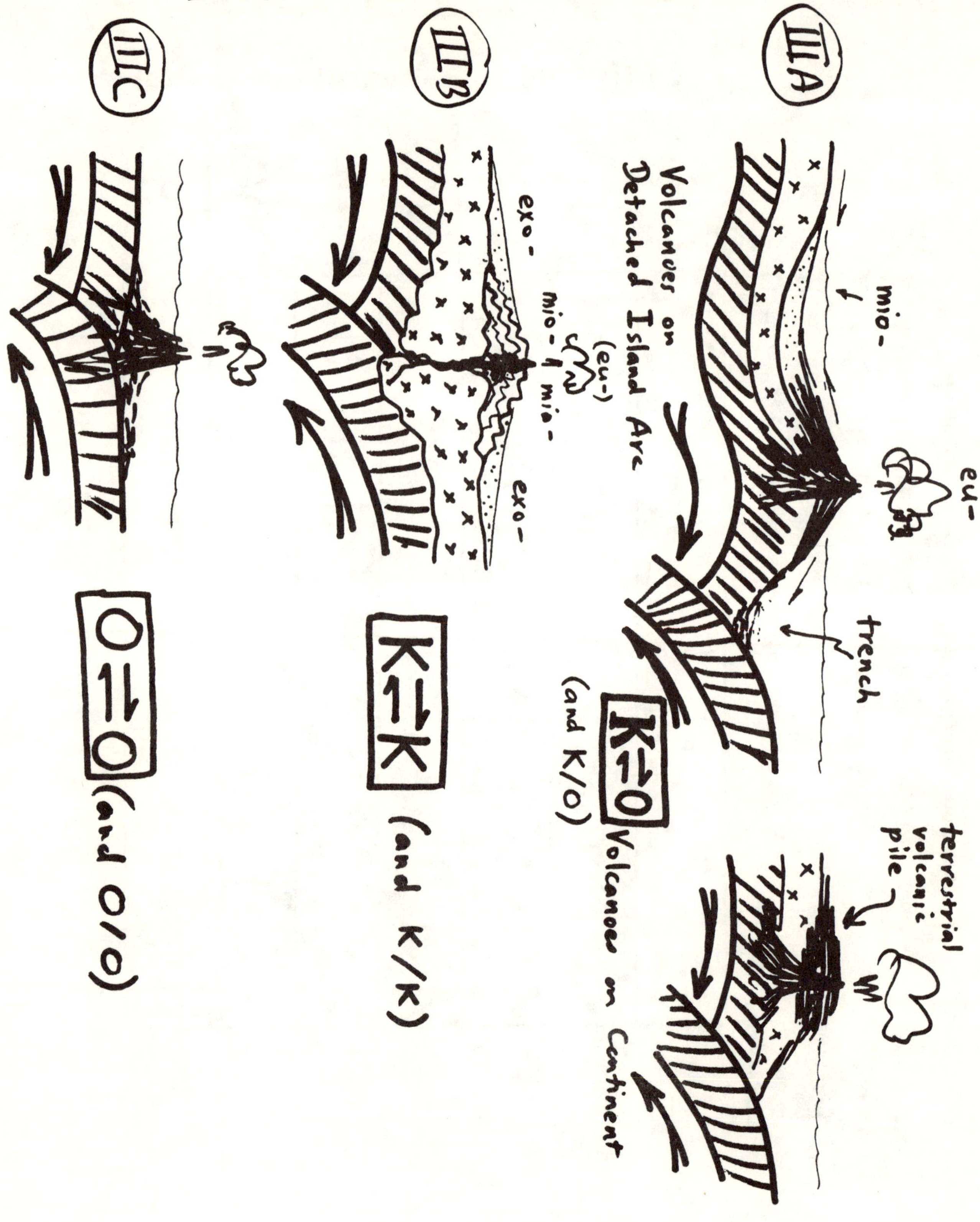

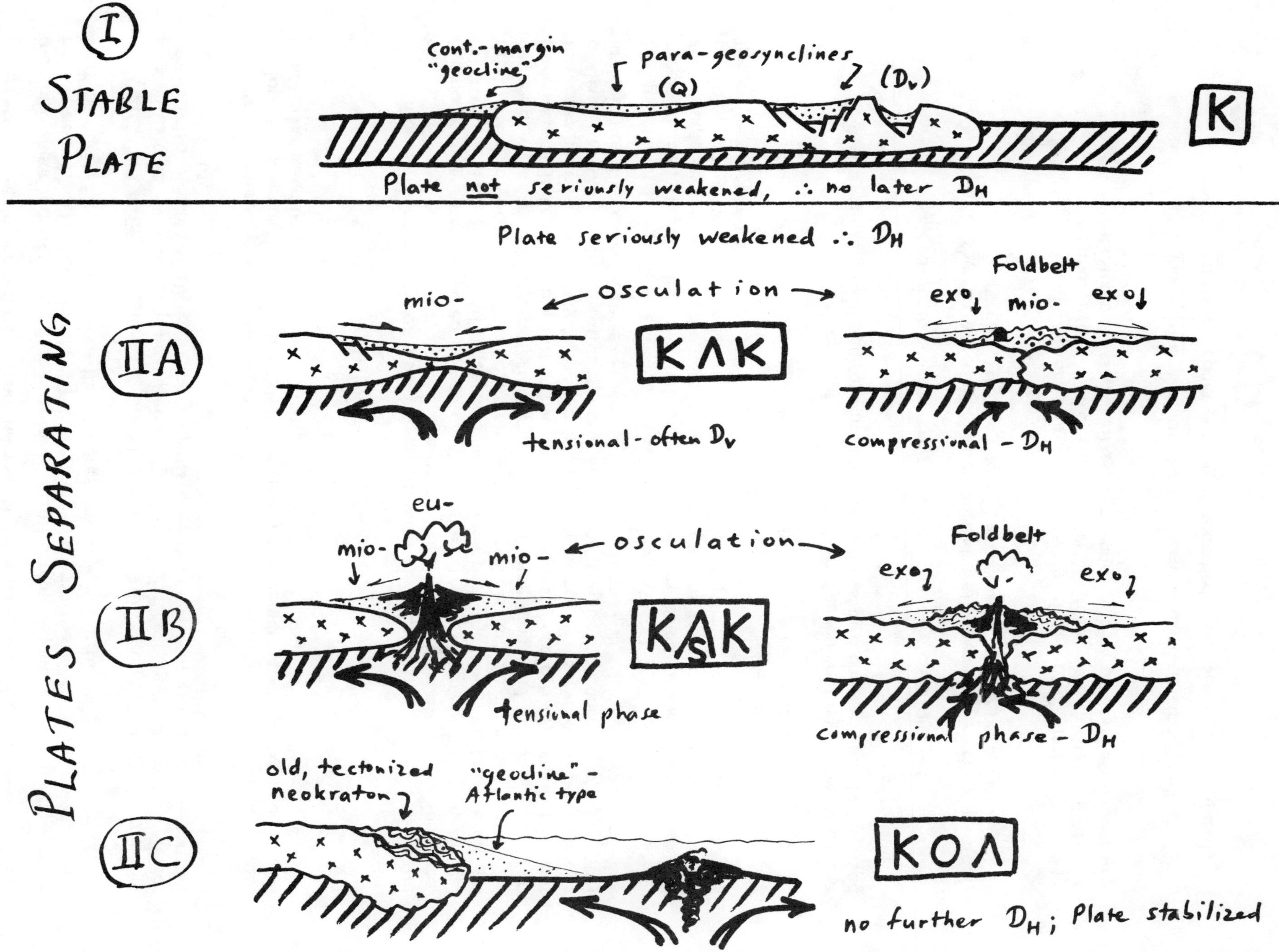
I
STABLE PLATE
cont.-margin "geocline"
para-geosynclines
(Q)
(D_V)
K
Plate not seriously weakened, ∴ no later D_H
PLATES SEPARATING
Plate seriously weakened ∴ D_H
IIA
mio-
osculation
KΛK
tensional - often D_V
Foldbelt
exo
mio-
exo
compressional - D_H
IIB
eu-
mio-
mio-
osculation
KΛK
S
tensional phase
Foldbelt
exo
exo
compressional phase - D_H
IIC
old, tectonized neokraton
"geocline" - Atlantic type
KOΛ
no further D_H; Plate stabilized

compression; resting or brief tensional phases (D_v) are possible. Two continental plates may also shear sideways (K/K) as in southern California. Again, volcanism appears to be suppressed. After continents are welded together and the scar tissue healed, a new split may arise in a new location--thus progressing back to I or II.

IIIC. PLATES COLLIDING, Ocean/Ocean ($0 \rightleftharpoons 0$). Forms chains of volcanic oceanic islands with surrounding aureole of volcanic and carbonate sediments. Shear motion also possible (0/0) as in the Cayman Trench. Often trenches do not contain much sediment.

In summary, most geosynclinal nomenclature is worthy of retention. Plate tectonics has simply provided the driving engine and revealed much better understanding of the reason behind the geosynclines.

Genetic Code for Sandstones

In deciphering genesis of a sandstone, we need to consider at least five elemental points. These are (1) tectonics of the source land, which controls erosion rate, geomorphology, etc.; (2) paleogeology of the source terrane, which controls mineralogy; (3) tectonics of the depositional site which controls thickness pattern, etc.; (4) depositional environment, which controls texture, sedimentary structures, etc.; and (5) degree of weathering of the material, which affects mineralogy and texture. All are to some degree linked, but the latter two are more or less "accidental," while the first three are linked more closely to the diastrophic framework. To serve as a shorthand, or at least an outline for meditation, any sandstone can be categorized by a symbol involving all five aspects; for example, D_v.G/p/C_f(w), or Q_k.S/s/L(n). If you don't like the shorthand, it can be written out in words. Krynine's "Holy Trinity" of sandstone types adopted the simple view that all five properties could be predicted from the source area tectonics; the scheme presented here emphasizes that all combinations are possible, though some are much more common than others. Now, to discuss each of these five aspects in turn.

I. Tectonics of the Source Area. Five endpoints are recognizable:

Q, <u>QUIESCENCE</u> or Stability; low relief; "peneplanation," slow sediment influx; mild epeirogenic warping and eustatic changes may be involved also.

Q_k, Kratonic Source--example, Canadian shield during Paleozoic time.

Q, source in neokraton, stabilized welt, peneplaned orogen, etc.--example, Atlantic coast Cretaceous (Schooley peneplane).

D, <u>DEFORMATION</u> of source area, with differential activity, folding, faulting, etc. More rapid detrital influx and higher relief. A gradational mess, but often two end-members can be recognized in a given region:

D_h, <u>Horizontal</u> Deformation--strongly asymmetrical folding, overthrusts, rocks become squashed and sheared, ("Alpinotype" deformation of Stille). Example: Appalachian welt during later Paleozoic. Often caused by collision of tectonic plates, or compression atop a subduction zone.

D_v, Vertical Deformation--block faulting, symmetrical folding, rocks merely jostled without much squashing ("Germanotype" deformation of Stille). Example: Basin Range province, horsts and grabens. Often caused by tensional movement accompanying separation of crustal plates, or rifting within a single plate.

R, REJUVENATION* of source area; source region uplifted without differential deformation, little jostling or squashing takes place; source lithology depends on past history. Transitional in "activity" between Q and D. Common on neokraton, e.g., Gulf and Atlantic Coastal Plain Tertiary.

X, EXTRUSIVE activity; volcanoes active during deposition of sediments, may be independent of any other tectonics, may happen almost any place at any time; thus associated with D_h,D_vR, and sometimes even Q. Probably most common in the island arc-subduction zone situation.

II. Paleogeology of the Source Area. There are four main divisions, namely G (granite, gneiss, and in general plutonic igneous); M (metamorphic rocks, schist to slate, metaquartzite); S (Sedimentary rocks, if dominantly a carbonate terrane, can be specially designated S_c); and X (extrusive, volcanic rocks) whether contemporary or not. Volcanoes active at the time of deposition of the beds in question would be shown DX.X/; erosion of old volcanic rocks by R.X/, D_h.X/, etc. Subequal mixtures can be shown by GM, MS, etc.; gradual changes during the unroofing of a region or burgeoning of volcanism can be shown M → G, S → X, etc. The rare basic plutonic source can be designated B.

III. Tectonic framework of the depositional site. Krynine adopted the view that thick prisms (p) were always associated with block-faulting (D_v); that sheet sands (s) were always associated with Quiescent kratons (Q_k); etc. It is undoubtedly true that these are very common associations, and perhaps they are the most common; but we must realize that all sediments do not fit such an ideal pattern, and should recognize existence of exceptions and not mentally sweep them under the rug. For example, sands can be washed off a stable kraton and accumulate in a violently subsiding neighboring trough. It is difficult to make a logical classification of basins, but here is an attempt--suggestions welcome on improving it.

s Sheet or blanket, very thin and widespread

s* Sheet-wedge, basically a sheet covering a large area, but greatly thickened at one side

b Basin, fairly equant shape, thin to moderately thick (e.g., Michigan Basin) b* Basin-wedge, equant depositional region, greatly thickened at one edge (e.g., so-called Catskill "delta")

p Prism, small area, often fault-bounded, extremely thick at one side with rapid thinning (e.g., Newark fault troughs).

g Geosyncline, long, narrow, thick trough such as a typical miogeosyncline

g* Geosyncline of extreme thickness (thickness like that of typical eugeosyncline--but no volcanism implied as that comes in another part of the symbol)

l Local patch, small, thin, often superficial or residual or local valley-fill

In the above, (*) means great thickness. This system is flexible so that account may be taken of units that may grade from one type of basin in one area to another type elsewhere--example, sb or sg.

IV. Dominant Depositional Environment

Tercier proposed a sort of super-environmental assemblage, upon which the following is based. A formation, though made up of rock bodies deposited in many micro-environments (e.g., tidal flat, estuary, lagoon, beach-dune, etc.) can be placed in a general mega-environmental assemblage as follows:

C Continental; flood plain, river channel, swamp, etc.

C_f Continental, alluvial fan.

C_l Continental, lacustrine.

C_d Continental dune-sheet (e.g., Saharan erg).

L Littoral: specifically mainly beach, beach-dune, some shallow marine "undaform"--basically shoreline deposits in areas not much influenced by riverine contributions.

P Paralic: floodplain and delta, coastal swamp and lagoon, estuarine, shallow marine; heavy influence of alluviation (e.g., Pennsylvanian of Appalachians). P_d: deltaic.

E Evaporitic: inland salt lake or marine evaporitic lagoons.

M Marine, generally shallow and of normal circulation. May be subdivided into M_u, undaform of Rich, wave-influenced generally coarser sediments; M_c, clinoform or slope deposits, if turbidites M_{ct}; M_f, fondoform of Rich, below significant stirring by wave action, generally muds; M_k, dominantly carbonate area with little detrital influx.

B Bathyal to Abyssal, slope and baisn deposits; clinoform and deep-water deposits; often "turbidites" (B_t).

C_g,M_g, etc.--Glacially influenced.

As before, mixtures can be shown CL, PM, etc.; gradual transition, M → B, P → L, etc.

V. Climatic Effects (when decipherable)

(w) Climate conducive to heavy weathering (hot or wet or vegetated).

(n) Climate not conducive to much weathering; if distinguishable, (n_d) too dry, (n_c) too cold, (n_v) lack of vegetation, etc.

All five aspects of the genetic code can be put together to give a systematic exposition of the origin of the sandstone, conglomerate, etc. The more "tectonic" influences are put first; depositional basin is set off by / / for ease in reading the formula; and the "accidental" influences, environment and climate, come last. Sic Semper Scientis!

Mineralogical Classification of Sandstones

The perfect classification of sandstones does not now and never will exist. Each of the dozens of classifications proposed have reflected the composer's personal bias and limited experience. The less one knows about a subject, the easier it is to cook up a logical classification that seems to have all problems answered. But no classification can be developed to answer all purposes and reflect all local peculiarities. Problems of classification have been thoroughly reviewed by McBride (1964 JSP), Klein (1963 GSA) and Okada (1971 JG) while Huckenholz (1963 JSP) and Dott (1964 JSP) have considered in particular the significance of arkose and graywacke.

The first English-language attempt to create a quantitative, comprehensive and geologically meaningful (as opposed to a random) Classification of sandstones was that of P. D. Krynine, who had developed essentially his complete system by about 1941, taught it starting in 1943, and published it in detail in 1948 (JG). His thinking on sandstone classification had apparently been strongly influenced by that of his teacher in Moscow, Shevtsov (see Folk and Ferm, 1966 JSP, for details and citations). Krynine stressed sandstone mineralogy as a powerful indicator of tectonics, and drew quantitative lines between five sandstone types, expressing composition by triangular diagrams based on poles of (1) quartz plus chert, (2) feldspar plus kaolin, and (3) micas (including MRF's and matrix together). After Fischer (1933), he was the first to express sandstone composition by a triangular diagram, though Pirsson's text used a ss-sh-ls sedimentary-rock triangle in 1915. My own ideas on mineralogical classification of sandstones were strongly influenced by Krynine; but the realization that an "arkose" with 35% feldspar could have a very wide range of textures necessitated some addition to the strictly mineralogical terminology, hence the idea of adding to the clan name a scale of textural maturity (1951 JSP).

The five-fold, textural maturity/mineralogical classification used in previous editions of this book (1954-1965) and published by the writer (1951 JSP, 1954 JG, 1956 JSP, 1960 JSP) was intended to be objective and descriptive, but the end-members of the sacred triangle were collected in three groups thought to be most indicative of provenance, generally following Krynine's philosophy. Hence quartz and chert were linked at the Q-pole, indicative of prolonged chemical/physical attack or else of derivation from an older sedimentary source terrane, the end-member rock type being the orthoquartzite. All feldspars and all plutonic and volcanic rock fragments were linked together at the F-pole, indicative of igneous source, the end-member type being

SANDSTONE TYPES:
RESULT OF PROVENANCE & TECTONICS

(MODIFIED BY: CLIMATE, DEPOSITIONAL ENVIRONMENT, DIAGENESIS, ETC.)

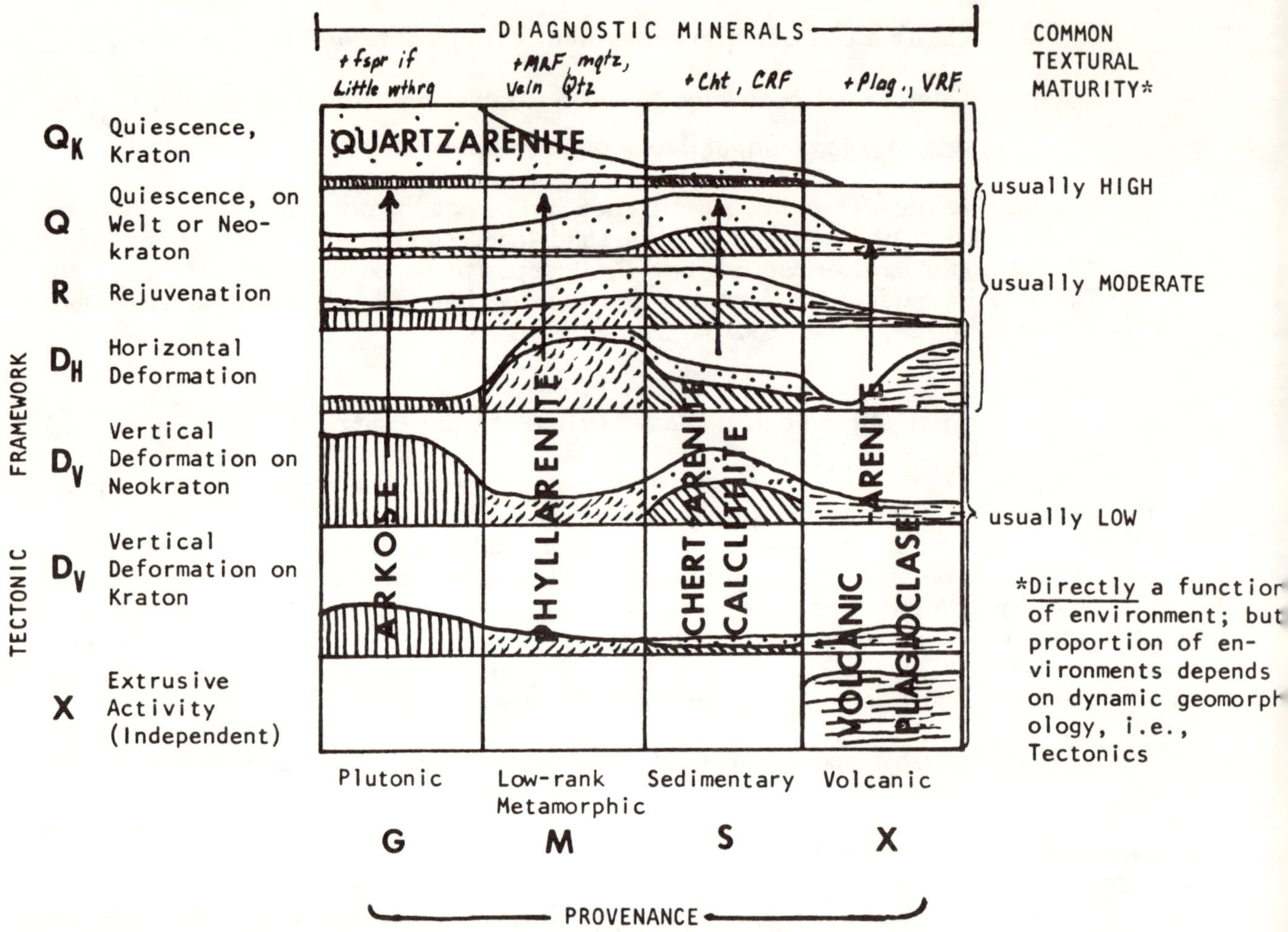

Thickness of lenses shows abundance (estimated) of rock types formed by each combination. Dotted = Quartzarenites.

TECTONIC ENDPOINTS

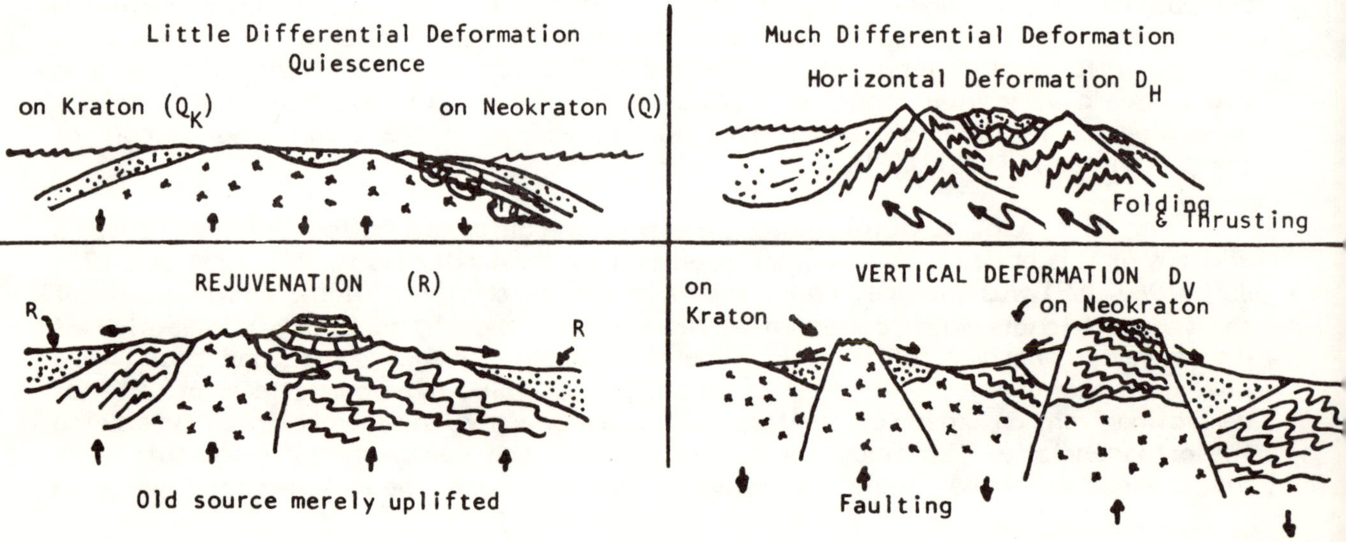

IDEALIZED GEOSYNCLINAL CYCLE IN MOBILE BELTS

Modified slightly from Bertrand, Shvetsov, Krynine, Stille, Kay, Pettijohn, etc. This cycle is a common one in many areas (Pettijohn, 1957) -- but some parts may be omitted or repeated; some parts may occur independently and without "system." As well as a "time" sequence, this may be a "space" sequence -- i.e., at any given time, part of the Kraton-mobile belt couple may be in one stage, and part may be in another.

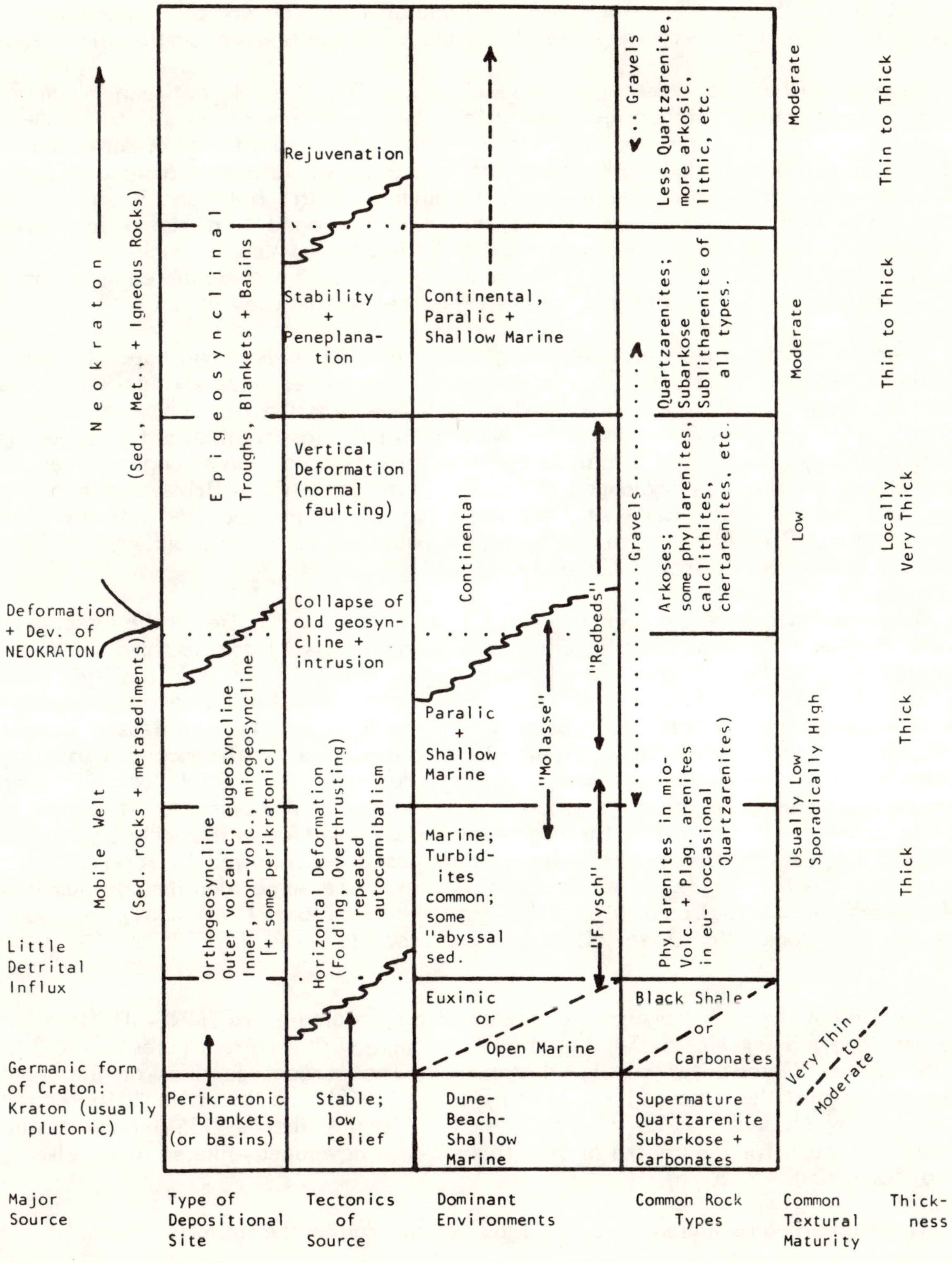

the arkose. Metamorphic rock fragments, micas, and metaquartzite fragments were linked together at the M-pole, indicative of metamorphic source, the end-member rock type being the "graywacke" (this is essentially the composition of the graywacke of Naumann, 1850 text). Later versions added a fourth pole for subaerially-eroded, much older carbonate rock fragments, defining a rock known as a "calclithite" with over 50 percent CRF's (1959 AAPG). Clays were eliminated from the compositional name (a major departure from the Krynine system), and placed in the textural part of the name.

Several difficulties arose with this scheme. The division between "ordinary quartz" (Q-pole) and "stretched metaquartzite" (M-pole) is subjective, difficult enough to determine with the microscope and certainly impossible in the field. A dark-colored sand grain might be a chert, a basalt fragment, or a piece of slate, thus assignable to Q-, F-, or M-poles; the distinction is rarely attainable in the field and is sometimes difficult even with the microscope. Chert itself shows a complete gradation into shale and siltstone, fine-grained metaquartzite, and silicified volcanic rock--and it is difficult to define these reproducibly. The word "graywacke" is encumbered by so many radically differing definitions that it has been rendered almost useless.

Therefore, after much soul-searching, the following revised and hopefully more workable sandstone classification has been used by the writer since spring 1966. It is hoped that those of us seriously interested in sandstone classification can, some decade, all get together and use some compromise system, so this classification is tentative and subject to change should such a miracle come to pass. I have been strongly influenced in this change of heart by prolonged discussions with Earle F. McBride, Keith A. W. Crook, and Harvey Blatt; many of their ideas have been incorporated into the new revised version. This new system has been now published formally [Folk Andrews and Lewis (1970), New Zd. J. Geol. & Geoph.].

The main difference between this and my previous classification is the lumping of all rock fragments (except plutonic ones, granite and gneiss) into the third or "rock-fragment" pole of the triangle, and the switching of chert--which is in reality a rock fragment--from the Q-pole to the RF-pole. Van Andel (1958 AAPG) also placed chert with rock fragments. Unfortunately, by putting all rock fragments regardless of genesis in one pole, the main triangle loses almost all its source-area significance; but splitting the rock fragments on the subordinate RF triangle restores much of this loss. Lumping of all rock fragments together to produce a "rock-fragment sandstone"--however it might be named--was essentially the approach used by Shvetsov in the early 1930's, and he termed the rock a "graywacke"; this system has been followed by later Russian petrographer (Shutov, 1965). Later workers who used the same idea include McElroy (1954 Austral, J. Sci., petromictic or lithic sandstone): Gilbert (1954 book); Pettijohn (1954 JG; 1957 book); Van Andel (1958 AAPG); Crook (1960 A. J. Sci.); and Mc Bride (1963 J. S. P.).

Splitting of the rock fragments by means of a subordinate VRF/MRF/SRF triangle is my own idea, (though Kossovskaya used a RF triangle with different poles in 1962) as are the names "Calclithite" and "phyllarenite." The percentage lines of the main triangle are almost the same as those used by Crook (1960), and the name "Litharenite" was coined by McBride (1963) who shortened it from Gilbert's (1954) term "lithic arenite." So much for the record of how this scheme developed--like so much else, a mass of borrowing.

To determine a rock clan under the 1966 system, follow the following steps.

I. Ignore the percentage of clay matrix (including micaceous hash finer than .03 mm), chemically-precipitated cements, glauconite, phosphates, fossils, heavy minerals, mica flakes, etc. Recalculate all essential constituents (those used in giving the sandstone clan name) to 100%, allotting them to one of the three following poles:

Q-pole: All types of quartz including metaquartzite (but not chert).

F-pole: all single feldspar (K or NaCa), plus granite and gneiss fragments (plutonic and coarse grained, deep-crustal rocks).

RF-pole: All other fine-grained rock fragments (supracrustal): chert, slate, schist, volcanics, limestone, sandstone, shale, etc.

These percentages will determine which one of the seven main rock clans the specimen falls into: If the specimen falls in the fields of Sublitharenite, Litharenite, or Feldspathic Litharenite, then proceed to II.

II. Recalculate all fine-grained Rock Fragments to 100% and plot on the RF triangle. This shows whether the rock is a volcanic-arenite, phyllarenite, or sediment-arenite (sedarenite??). If the latter, then go to III.

III. Recalculate all Sedimentary Rock Fragments to 100% and plot on the SRF triangle. This determines whether the rock is a chert-arenite, calclithite, sandstone-arenite or shale-arenite.

An alternative, simpler procedure would be to simply forget the daughter triangles and name the rock according to the most abundant rock fragment.

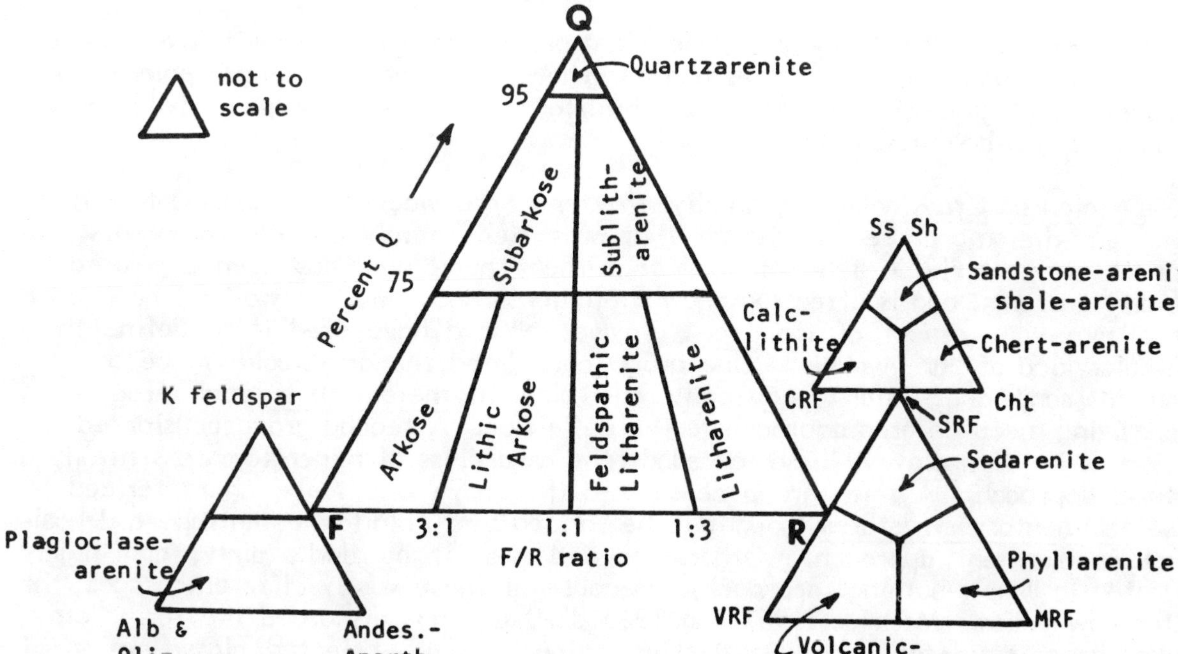

Notes: The garbage-can words "Litharenite," "Sublitharenite," etc. should only be used as group terms, or when (as in many megascopic descriptions) it is not possible to identify the specific rock fragments. The more exact terms "phyllarenite," "feldspathic phyllarenite," "chert-arenite" should be used if at all possible. Instead of "Sublitharenite" one can use "Subphyllarenite" (or Quartzose Phyllarenite), "Quartzose Volcanic Arenite," etc. If more specific information is needed, one can use "rhyolite-arenite," "andesite-arenite," etc. If rock fragments are subequally mixed, perhaps "polylitharenite" is a useful term.

For rocks that do not logically fit on this triangle, which is intended for volumetrically important sandstones, such names as "glauconite-arenite," "phopsphate-arenite," "augite-arenite," "gypsum-arenite" are perfectly satisfatory.

The unfortunate characteristic of this plot is that it is no longer possible to show provenance by the position of one dot on the main triangle alone. Of course, two dots (one on the main triangle, one on the RF triangle) do pretty well indicate provenance, but this becomes messy with a lot of samples. The only way to indicate provenance satisfactorily is by a separate graph. One good way to show this is by a star diagram, such as this:

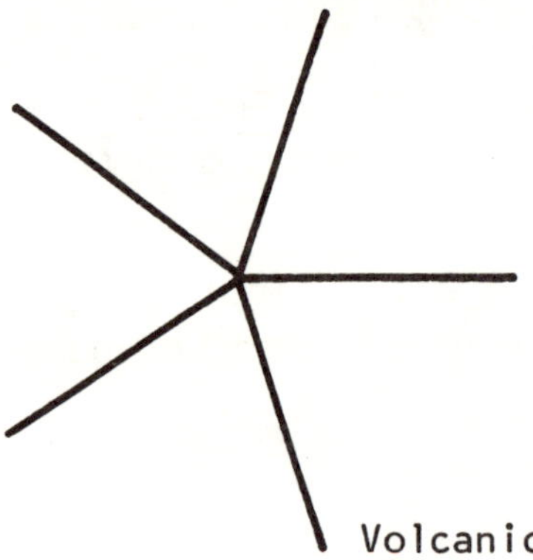

Here, at the "sedimentary" pole, one might place together quartz with overgrowths, chert, super-rounded quartz, limestone fragments; at the "Volcanic" pole Quartz phenocrysts, plagioclase and VRF's and biotite; at the "Metamorphic" pole, meta-quartzite, schist and muscovite.

Opinion on Graywacke. Originally, the term "graywacke" was applied to a hard, dark, semi-metamorphosed sandstone that was rich in mixed rock fragments and chloritic clay matrix, a type of rock that many now think has been deposited by turbidity currents, and is often somehow metamorphosed. Modern workers have each seized some one aspect of the type "graywacke" and have used it to define their particular kind of "graywacke." One group considered the graywacke to be a rock-fragment sandstone, later specifically one rich in metamorphic rock fragments, emphasizing a source-area approach to the definition. A second group considered the graywacke to be a clayey, ill-sorted sandstone regardless of mineralogy, essentially a textural approach. A third group considered the graywacke to be characterized by those sedimentary structures thought to be formed by turbidites, essentially a depositional environment approach. Others insisted that it be dark, dirty, and highly indurated (a "know-nothing" approach). Because of these vastly different usages, the writer now follows McBride (1962 and 1963 JSP) and has discarded the term "graywacke" from any seat in a quantitative, mineralogically-oriented classification of sandstones. It should revert to a status like that of "trap rock" as a very loose, field term--essentially a very hard, ugly, dirty, dark rock that you can't tell much about in

DESCRIPTIVE (MORPHOLOGICAL)

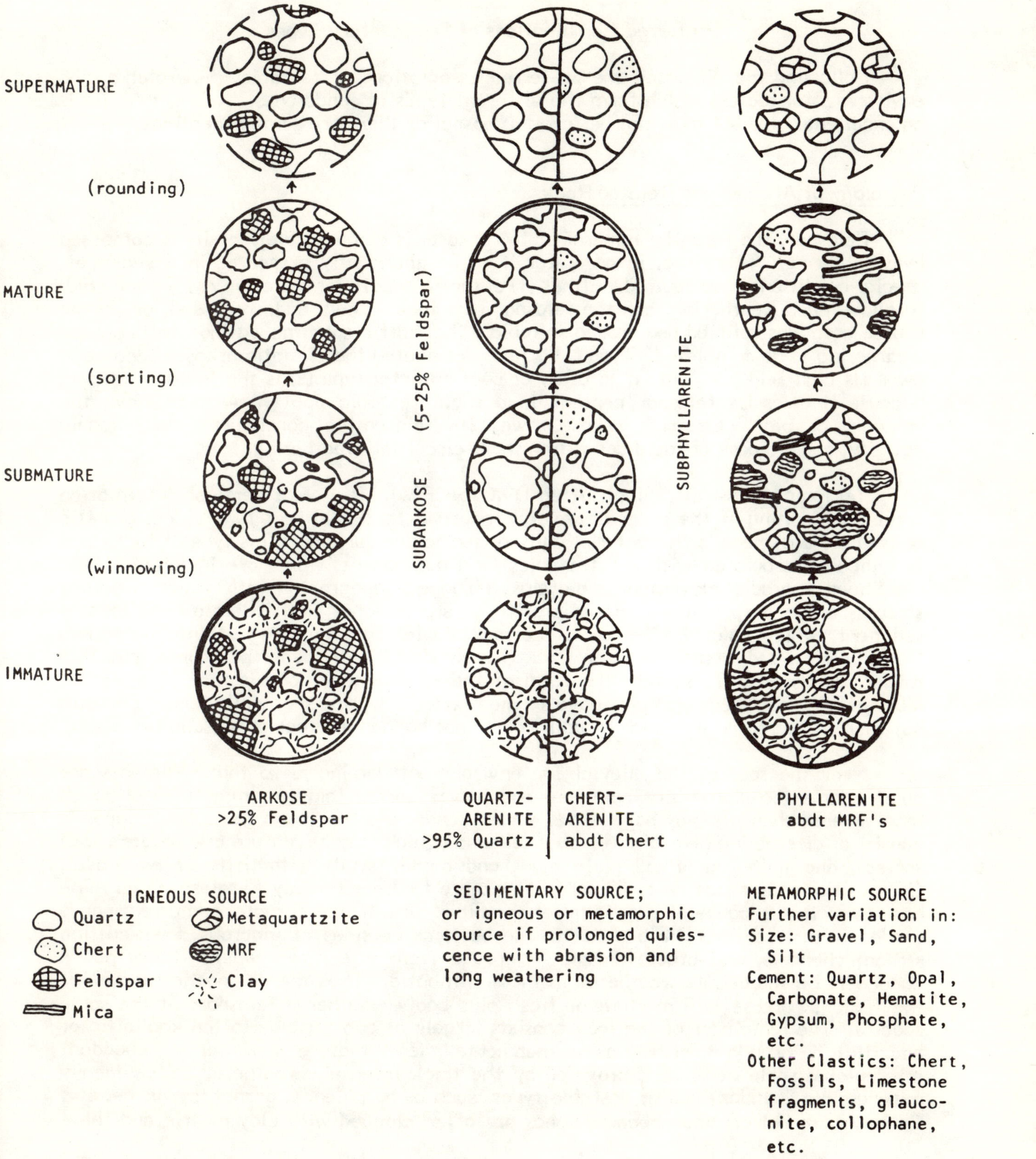

the field. McBride wishes to define it compositionally as having over 15% chlorite/sericite matrix, over 10% unstable rock fragments, and over 5% feldspar. It is a special, polyvariant term in which hand-specimen properties are more important than the precise composition or texture.

Preferred Combinations of Sedimentary Types

Following are discussed the preferred associations of tectonic, mineralogic, and environmental factors which form the different types of sandstones. Again, this list is an excellent guide but must not be taken as gospel as there are some exceptions.

Petrology of Arkoses and Related Rocks

Feldspar, the essential mineral of the arkose, is comparatively easily decomposed by weathering. Therefore, in order for it to be abundantly presented in a sediment, special conditions are required. Either (1) climatic conditions were too dry (or too cold) to allow much weathering, and the feldspar remained largely undecomposed, or (2) the source area was uplifted and eroded so rapidly that sufficient time was not available for weathering to be completed. Type (1) is here designated the climatic arkose, because it owes its high feldspar content to a dry or cool climate; type (2) is the tectonic arkose, because it owes its feldspar content to a rapid tectonic uplift. A third type, the volcanic or plagioclase-rich arkose, owes its feldspathic constituents to volcanic activity, again a case of rapid deposition short-circuiting weathering.

Tectonic Arkose (Krynine), often Dv.G/p/CF(w). If violent vertical deformation and block-faulting of the order of several thousand to 20,000 feet take place, granitic or gneissic basement is characteristically raised to the surface. This type of tectonism is typically associated with KΛ tectonics (and aulacogens, which are incipient KΛK). Continued sporadic movement maintains a rugged topography, into which vigorous streams carve deep valleys. Rapid erosion on steep slopes contributes a huge load of sediment, which quickly piles up as coarse alluvial fans passing out into aggrading fluvial plains; sediments are very thick (thousands of feet) near the source, but thin rather rapidly outward so that the form of the seidmentary body is that of a wedge. Most of the described examples have formed under a warm, seasonally humid climate, but there is no reason why such an uplift could not happen in areas of dry climate also.

Near the source, the alluvial fan environment dominates so that sediments are quiet coarse (conglomerates common) and poorly sorted (clayey, immature) although local stream channels may have better sorted sands. Farther from the source, deposits consist of alternating beds of river-channel sand (which may be mature arkose, i.e., well sorted), and immature flood plain muds and muddy sands with little or no gravel. Composition may range up to 50 percent or more feldspar (mainly K-feldspar); Krynine has shown that because of rugged relief combined with humid climate, some feldspar is eroded from fresh granitic bedrock in the bottoms or sides of vigorous down-cutting stream channels, and other feldspar is eroded from interfluves where it has had a chance to be thoroughly weathered in the soil mantle. Thus the diagnostic marker of the tectonic arkose is a mixture of fresh plus badly weathered feldspars of the same species. The remainder of the rock consists largely of quartz; clay (often kaolinitic or bauxitic) is abundant either as a matrix to the immature sands or as abundant intercalated shale beds; it is provided by the thick intensely weathered soils. Heavy minerals are abundant and of unstable types, such as hornblende, garnet, pyroxene, etc. Cements are not common because sands are often clogged with clay matrix and these

INCREASING MATURITY →

Maturity may increase by more tectonic stability, which in turn means lowered physiographic relief, decreased rate of subsidence, decreased rate of deposition and consequent increase of time available for chemical weathering (if any) and abrasion. Maturity is also increased by any of the above factors operating independently, or by a change in environment. It also increases with distance from source.

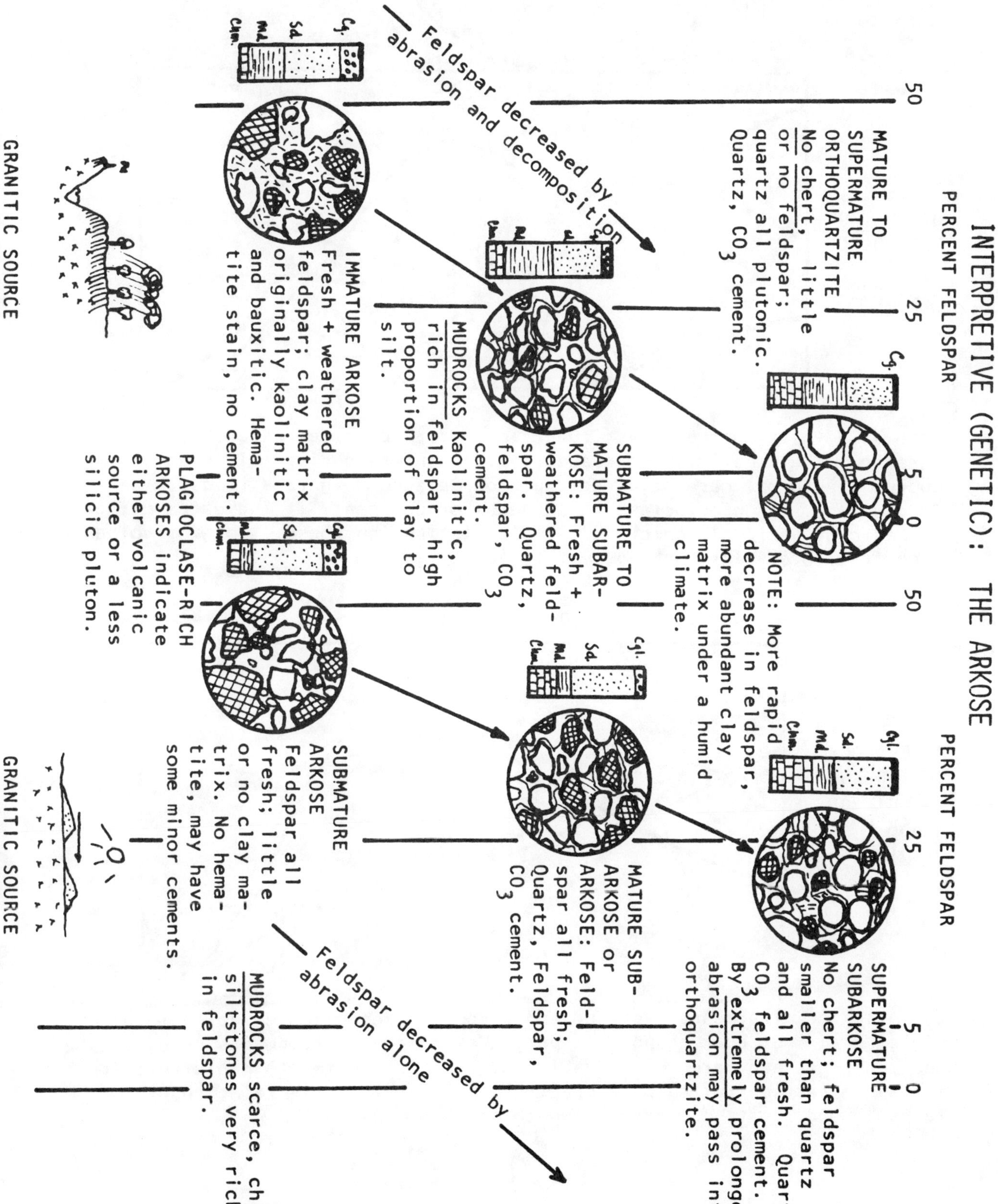

MINERALOGY: THE ARKOSE

MINERALOGY CONTROLLED BY SOURCE AREA

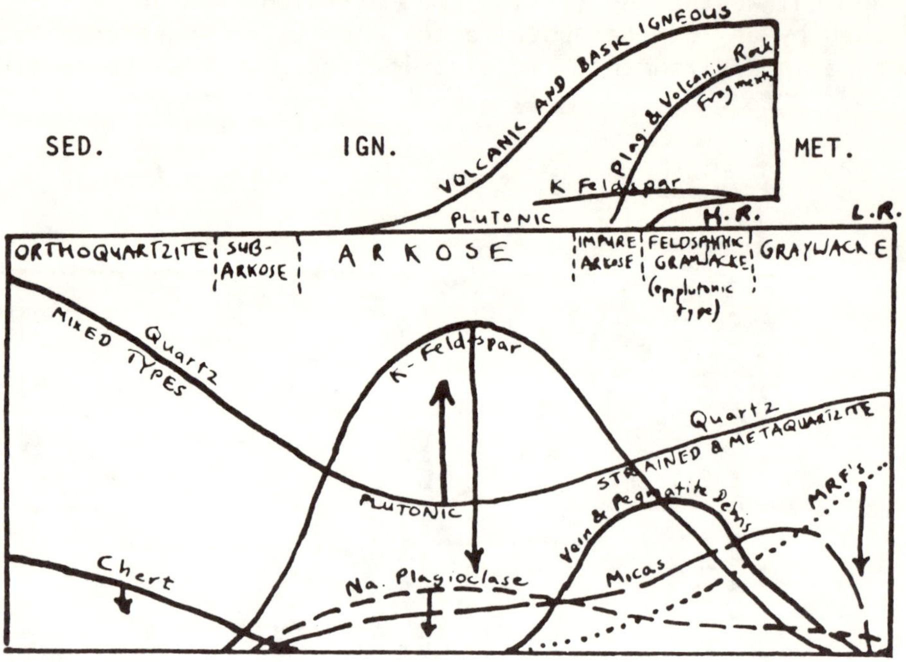

Arrows indicate whether constituent is increased or decreased by prolonged abrasion. Feldspar also decreases rapidly on weathering. In diagram above, the normal care of an acid pluton is shown, where K feldspar >> Plagioclase. If plagioclase is subequal or dominant, it indicates either volcanic or basic plutonic source. To distinguish these, look for volcanic rock fragments, glass, phenocrysts, etc.

MINERALOGY CONTROLLED BY GRAIN SIZE

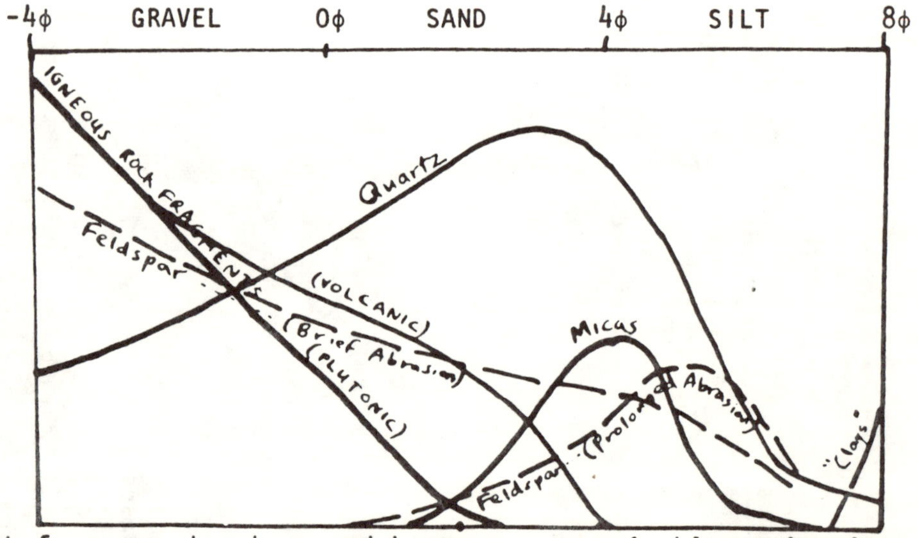

Plutonic rock fragments break up and become unrecognizable, going into individual grains of quartz and feldspar. Volcanic rock fragments being tougher and finer-grained, are recognizable to much finer sizes than plutonic fragments. Micas controlled by hydraulic factors. The size relationship of feldspar and quartz depends on length of abrasion.

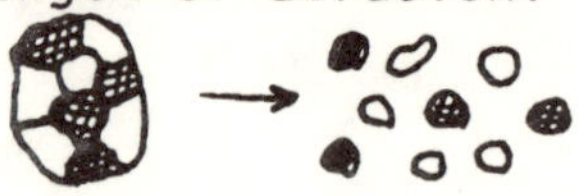

are usually continental sediments. These arkoses are usually red because of pink feldspar and a red, hematite-stained clay matrix either derived from the red soil developed under warm, humid climates; Walker believes that red color may form also in dry climates by intrastratal oxidation of heavy minerals.

Variants from this polar type are fairly common. Many times the block faulting is not of enough magnitude to expose "pure" granite or gneiss; areas of schist, metaquartzite or veins then contribute considerable strained and hydrothermal quartz, metamorphic rock fragments, micas, and typical metamorphic heavy minerals to the arkose and it consequently grades into an impure arkose. If faulting is of still smaller magnitude or a thick sedimentary cover is present, this blanket may not be completely eaten through, and reworked grains of older sandstones, cherts, and limestones are contributed; the rock may even grade into a calclithite if carbonate detritus dominates. Infrequently, in place of granite a more basic pluton furnishes detritus; here the feldspar will be dominantly plagioclase. If climate is dry instead of humid, there will be much less clay, the red hematitic matrix will be absent, and the feldspar will all be pretty fresh. Block faulting bordering the sea (as in California) produces marine arkoses with different textural features, and some of these are turbidites.

Climatic arkose, often Qk.G/s/CD(n). Under some situations a granitic land area becomes peneplaned during a period of tectonic quiesence. Such may happen either to areas that have been stable for a very long time (e.g., Canadian shield), or to areas that have been uplifted, undergone deep erosion, and then were planed down on cessation of uplift. If the climate is humid, the feldspar will all be decomposed because of the slow rate of erosion on such low-relief topography, and quartzarenites will result. But if the climate is dry very little chemical decay of feldspar will take place and supermature arkose results. Feldspar is eliminated by abrasion alone, because of its inferior hardness. Because of tectonic quiescence and low topography, beaches and dunes comprise the bulk of the sedimentary environments. No areas are subsiding or being uplifted greatly, consequently, a sheet of sand is spread smoothly and evenly by broadly transgressive or regressive seas; the very slow influx of detritus permits uniform spreading of the sediment. Thus sands are of the blanket type. Sediments are supermature because of the high abrasion and sorting efficiency of the beach-dune environment and the great length of time available under such slow conditions of erosion and deposition. The sediments are superbly rounded fine to very fine sands, with the feldspar fresh and usually finer than the quartz (Folk). Feldspar content can range up to 40 percent or more; fine-grained beds may carry abundant feldspar and intercalated coarser ones very little. Common quartz is the only other major constituent; heavy minerals are almost soleiy rounded tourmaline, zircon and occasionally garnet because little else can stand such prolonged and effective abrasion. Cements may be quartz, carbonates, evaporites, or quite frequently authigenic feldspar overgrowths. There is little associated shale section because of the lack of soil in the source area. The color is white, so that in the field they are very commonly mistaken for quartzarenites. Close study usually reveals whitish or light gray specks of feldspar. With prolonged abrasion, these rocks pass into subarkose (the most common type of climatic "arkose" is really a subarkose) or even into quartzarenites.

Volcanic Arkose and Plagioclase Arkose, often X.X/b/C. One type of plagioclase arkose (that eroded from a basic pluton) has already been discussed. Most plagioclase arkoses and subarkoses, however, are formed when volcanic activity contributes material to rapidly depositing sediments. Deposition must be rapid because of the very unstable nature of basic volcanic materials. If the volcanic materials are chiefly bits of aphanitic ground-mass, then the term "volcanic-arenite" (or, if specific identification is possible, "andesite-arenite," "rhyolite-arenite," "basalt arenitic," etc.) should be used; if it is chiefly plagioclase, use "plagioclase-arenite" as the clan name. These rocks imply unstable tectonic conditions and rapid deposition of volcanic materials plus other diluting terrigenous detritus in fans, floodplains and river channels; hence maturity is immature to mature. Grain size varies from conglomerates to silts. Volcanic biotite is a common accessory as is apatite, together with basaltic hornblende and other more common ferromagnesians. Sometimes these constitutents may be so abundant that the rock can be called "hornblende-arenite," etc. Clays may be chloritic or montmorillonitic, occasionally grown as authigenic fringes; a chloritic matrix is very common if the rock fragments are basic. Cements, if any, are usually calcite, opal or zeolite. Volcanic-arenites are usually dark colored in the field (if derived from andesites, or basalts); plagioclase-arenites are light gray. If volcanic glass is a major component (implying a siliceous magma), the rock may be classed as a "vitric-arenite"; these rocks are light gray and, on the Texas Gulf coast, contain much idiomorphic zircon (Callender).

Petrology of Phyllarenites and Related Rocks

Metamorphic rock fragments, the chief essential constituent of the phyllarenite, are soft and quite susceptible to abrasion. Therefore their abundance in sand (consequently the clan designation--phyllarenite vs. subphyllarenite vs. quartzarenite) varies quite enormously with environment of deposition, distance from source, and tectonic activity. The former term for this sandstone type was "graywacke," now abandoned as completely prostituted. The term "phyllarenite," coined in 1965, was suggested by an older term, "schist-arenite" as used by Knopf in the 1930's. "Phyllarenite" implies that the major constituents are foliated metamorphic rock fragments: slate, phyllite and schist.

Krynine (unpublished notes) divided his "graywackes" into "Drowned Graywackes" (mio- and eugeosynclinal) and "Floating Graywackes," mainly post-orogenic, with transport direction outward from the kraton instead of inward from an offshore welt. This is the basis for the classification used here: (I) Orthogeosynclinal deposits, including (IA) Miogeosynclinal Phyllarenite, generally the product of crustal plate juncture, KΛK in compression phases, or K K, and (IB) Eugeosynclinal rocks (composition variable, volcanic and metamorphic); and (II) Rejuvenation Phyllarenite. Orthogeosynclinal phyllarenites tend to be deposited in long, narrow geosynclines along the continental margin with detritus derived from a welt beyond the margin and moving dominantly inward, toward the kraton (as in the Appalachian geosyncline). Near the continent, the geosyncline subsides less rapidly (miogeosyncline) to produce "normal" phyllarenites lacking feldspar, while the outer geosyncline subsides more rapidly and also receives volcanic contributions (eugeosyncline of Stille) to produce rocks that range widely in composition from phyllarenite to volcanic-arenite to plagioclase-arenite. Rejuvenation Phyllarenites can be produced any place; an ancient metamorphic terrane is simply uplifted (without further squashing) to produce a rejuvenated source area, and they need have no connection at all with orthogeosynclines. A fine example (first pointed out by H. Blatt) is the southern Appalachians, an area metamorphosed in the Paleozoic but which provided a flood of Rejuvenation Phyllarenites that spread all the way to Texas in the Eocene (Todd, AAPG 1957) and which, in the year 1979, is still providing phyllarenite sands to the local rivers.

Miogeosynclinal Phyllarenite, often D_h. MS/g/PCM(w). Intense horizontal deformation on the proto-continental margin raises a welt or series of welts of considerable relief, separated from the cratonal mass by a rather rapidly subsiding geosyncline. The intense folding (sometimes isoclinal) and major overthrusting that accompany horizontal deformation often convert older shales, sandstones and other rocks into low-rank

INTERPRETIVE (GENETIC) - PHYLLARENITE

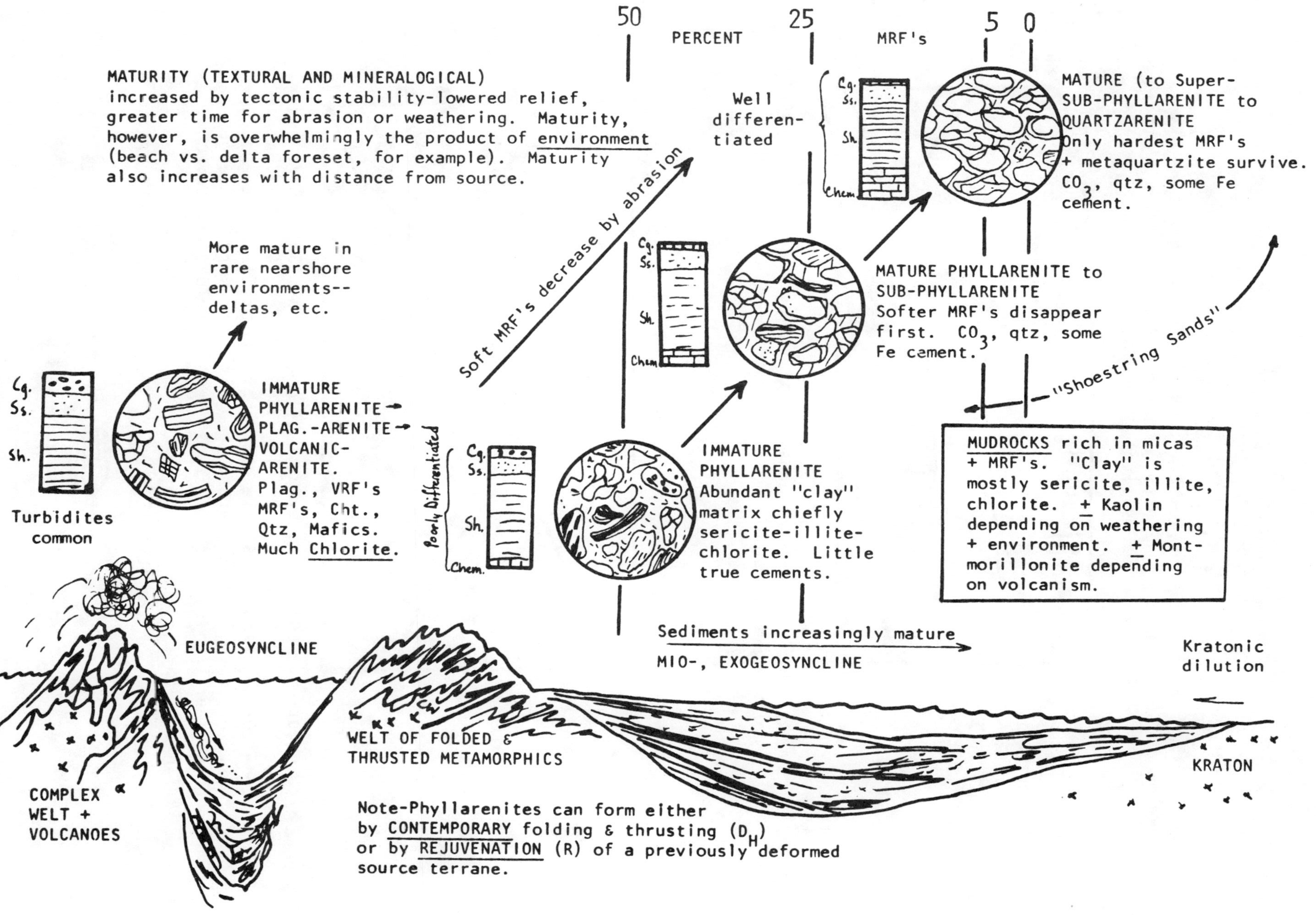

MINERALOGY: THE PHYLLARENITE

MINERALOGY
CONTROLLED
BY
SOURCE
AREA

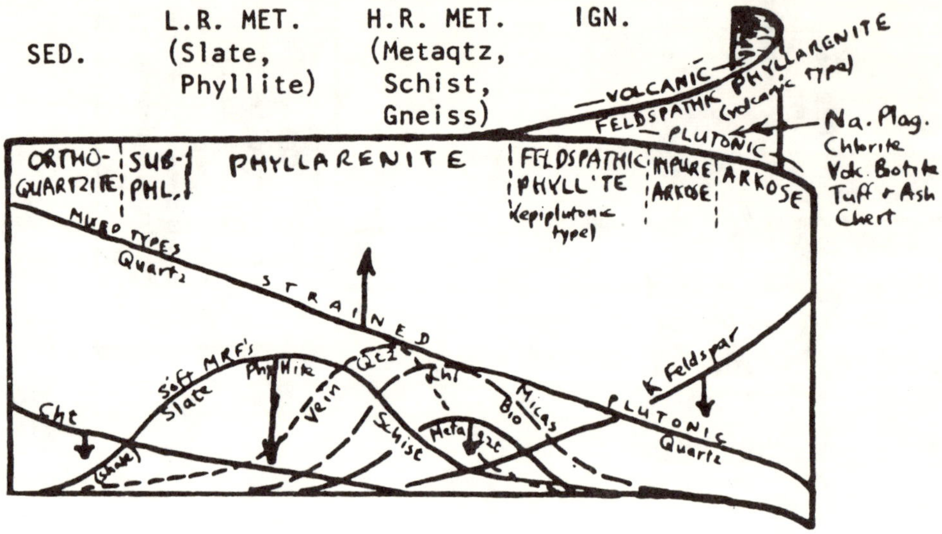

Arrows indicate whether constituent is increased or decreased by prolonged abrasion.

MINERALOGY
CONTROLLED
BY
GRAIN
SIZE

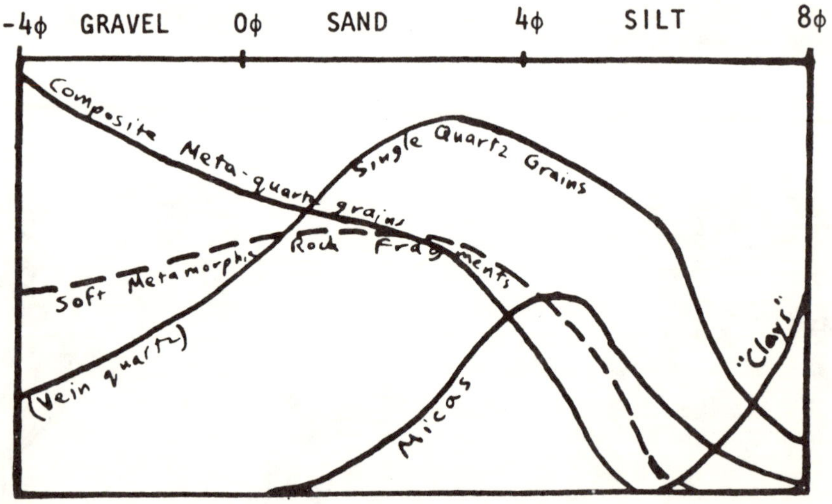

Mica occurrence controlled by hydraulic factors. Soft MRF's and composite quartz break up into clays and single quartz grains and become unrecognizable. Single quartz grains at first increase by breakup of composite metaquartz, then diminish; very little quartz occurs finer than 6φ - 7φ. It retains undulose extinction to 5φ - 6φ.

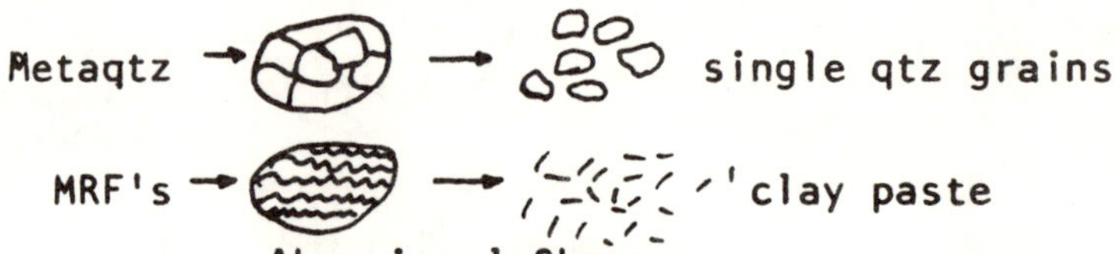

Abrasional Changes

metamorphics such as metaquartzite, slate, phyllite and schist. Furthermore, uplift attendant on the rise of the welt brings deeper crustal areas to the surface, and some intrusion and volcanism may occur. But dominantly the source area consists of low-rank metamorphic rocks, carved into considerable relief as it is continually deformed and raised. The resulting great flood of detritus is dumped rapidly in a series of coalescing alluvial fans, flood plains, river channels, large and small deltas, coastal loci such as beaches, swamps, lagoons, and estuaries, and neritic environments. Since this type of deformation produces linear welts, sediments accumulate in a linear trough which may be quite thick in the middle and thins to both sides. Climatic evidence is difficult to obtain, but it seems likely that a rugged welt surrounded on all sides by seas would usually have considerable rainfall.

Because of the softness of the slate and phyllite fragments, environment of deposition has a tremendous influence on the composition of the phyllarenite. Near the source where sediments are deposited rapidly, sands are immature--simply choked by the flood of fine-grained, clay-mica paste which is produced by abrasion of the low-rank micaceous source rocks (some "matrix" may readily be squashed MRF's--it is hard to be sure). These sands may contain up to 70 percent of metamorphic rock fragments (MRF's). Traced farther away, the softer fragments are abraded out and the sands become more mature and may pass into subphyllarenites, although mature phyllarenites are common in high-energy loci such as river channels. Rapid abrasion of slate and phyllite fragments produces great volumes of clay-mica mush, which are winnowed out of the sands to produce thick shale section. At intermediate distances from the source, environment plays the strongest role; beach sands (although not abundant in such a rapidly subsiding basin) may have lost all their soft MRF's and contain only stretched metaquartzite as a reminder of their origin; some of these may pass into quartzarenites or supermature subphyllarenites if a brief period of quiescence stabilizes the beach line. But fluvial sands at the very same distance from the source may still contain considerable MRF's, and floodplain or deltaic sediments may be extremely rich in them. Stratigraphic sections that have been deposited in this type of coastal region consist chiefly of shales (marine, estuarine, lagoonal, or floodplain), alternating with lenticular beds of texturally mature sands ranging from subphyllarenites to phyllarenites. But sand beds and clay beds are cleanly differentiated, which is not the case nearer the source. Far from the welt, most of the softer MRF's disappear and only metaquartzite and strained quartz remain. Dilution with detritus contributed from the kratonal side of the geosyncline also occurs. Texturally, the phyllarenites average fine to very fine sand (the finest-grained clan, on the average) because much of the source material breaks down into fine particles, and the bulk of the phyllarenites are deposited in low energy environments like deltas or sluggish rivers. Maturity depends on the environment of deposition but is usually rather low. Mineralogically they contain abundant metamorphic quartz, metamorphic rock fragments, and micas (especially muscovite). Shales and silts are especially micaceous. Pebbles are not uncommon in high-energy loci and consist of vein quartz and metaquartzite. These rocks often contain some chert. Potash feldspar is usually lacking or present only in very small amounts. Clays are chiefly sericite, illite, and chlorite because they are derived from breakdown of MRF's, but kaolin (sometimes authigenic) is common from heavily weathered source areas. Heavy minerals are simple (hornblende, zircon, tourmaline) because metamorphism in the source area is still of the low rank type which produces few fancy minerals. Winnowed sands are cemented with carbonates or quartz. These phyllarenites are usually gray or gray-green (because of the dark color of particles of slate, etc.) but staining by hematite may give red or purple colors. Tertiary phyllarenites of Texas contain light-colored MRF's which make the rock appear to be full of whitish flour specks.

Variants are common. It has already been mentioned that under more-than-usual abrasion, subphyllarenite or quartzarenites develop with few MRF's and some meta-quartzite grains; in hand specimen these look like clean, white, well-sorted quartz sandstones but they may contain a few scattered black or white specks. These rocks may occur either far from the source area or along local, temporary strand lines nearer the source. During brief periods of quiescence which occasionally interrupt geosynclinal subsidence, such sands may spread as blankets entirely across the basin; frequently these sands are supermature, and although usually classifiable as subphyllarenites, intensive abrasion may make them pass into quartzarenites. Often these sands contain pebbles of vein quartz or metal quartzite. Continental deposition under a humid climate in the source area may produce red phyllarenites, developed if uplift overbalances the rate of subsidence so that detritus spills into the basin faster than sinking can take care of it, and a great delta (exogeosyncline) is built up. Uplift or erosion may be so great (especially, late in the history of the orogeny) that higher-rank metamorphics or even the upper fringe of the granitized zone are exposed; potash feldspar is then contributed in small amounts and the rock grades into a feldspathic subphyllarenite. If any volcanic materials enter the sediment, small amounts of plagioclase, volcanic rock fragments, biotite, apatite, and montmorillonitic clay may appear. These, however, are more typical of the eugeosynclinal phyllarenites.

Eugeosynclinal rocks, often $D_h X$. MX/g*/B_t. These appear to be mainly products of simatic activity along trenches and subduction zones, $K\underset{s}{\wedge}K$ or $K \rightleftharpoons O$. Accumulated in the very rapid subsiding troughs or in the arc-trench gap with accompanying extensive volcanism, they are very poorly sorted and immature, and deposits may consist of interbedded muddy conglomerate, muddy sandstone, deep water black shale and radiolarian chert. Shoreline facies are present but rare (Dott). Turbidity currents operate par excellence in this environment. Basically, they are composed of a gamut of metamorphic debris--slates to schists and gneisses to which is added vein quartz, chert, a little K-feldspar from granitized zones and pegmatites and, most important, plagioclase and volcanic rock fragments. Compositionally, they show a very wide range; many could be called "feldspathic polylitharenites" because they contain such a variety of rock fragments--but mostly they range from phyllarenite through volcanic-arenite. In some areas, plagioclase-arenites are common. Typically they are made very hard by a blackish-green matrix of chlorite--partly caused by basic volcanic material, partly of metamorphic origin; sometimes zeolite cement appears. Inasmuch as many of these rocks have suffered later metamorphism (forming as they do in such active tectonic zones) their original characteristics are obscured, and burial metamorphism is common (epidote, pumpellyite, etc.). Some so-called eugeosynclinal rocks may be simply tectonic arkoses metamorphosed to chlorite grade (W. Harris).

Rejuvenation Phyllarenite, often R. M/s/CP. These rocks are produced by uplift (without further deformation) of older metamorphic sources (often neokratons). They may range from local valley fills to fairly large basins, to coastal-plain geosynclines like the Gulf and Atlantic coast Tertiary of the USA. They tend to be low in MRF's (actually subphyllarenites) because sources are mixed (older sediments, granites, etc.), and the ratio between uplift and abrasion plus weathering is low, so that a lot of the weak MRF's are removed. It appears that most of these rocks, if continental, are of subphyllarenite composition with some chert, feldspar, and other contaminants; littoral or marine equivalents tend to be quartzarenites instead. Heavy minerals may be very complex because of mixed source.

Petrology of Quartzarenites

Krynine's original scheme of the 1940's used the term "quartzite" for sandstones made of quartz ± chert; as this was confused with metamorphic rock, Pettijohn (1949) used instead "orthoquartzite" (this term was apparently coined by Tieje (1921, JG). Even so, this still misled the average geologist into thinking metamorphically, so gradually the term quartzarenite (Gilbert; McBride) has superceded it. Though less euphonious and longer, it will be used throughout this book.

Quartzarenites originate in diverse ways, but their high degreee of purity usually requires a long period of abrasion and/or weathering to remove the less stable feldspar, metamorphic rock fragments, etc. Quartzarenites may be developed directly from a granite or granitic gneiss under humid conditions if tectonic stability is sufficiently prolonged (say of the order of one-fourth to one-eighth of a geologic period). If the climate is arid, however, a quartzarenite may be made directly from a granite only under an exceptional and very prolonged period of tectonic stability coupled with intensive beach and dune action, because the feldspar must be removed by abrasion alone, unassisted by much chemical decay. Quartzarenites may be derived from metamorphic sources, but this also requires rather prolonged stability (a fraction of a geologic period) to remove all the metamorphic rock fragments. Some are made largely of vein quartz; these also require prolonged abrasion. If the source area consists of older sediments, however, a quartzarenite may be developed very rapidly because the source material is already quite enriched in quartz. Hence quartzarenites derived from reworked sediments require no period of stability, in fact many of them are orogenic sediments.

Plutonic source, humid climate, frequently Q_k. G/s/L(w). To obtain a rock that has less than 5 percent feldspar directly from a source granite-gneiss containing 70 percent feldspar requires prolonged abrasive or weathering activity of the most intensive kind. Such intense work takes place only on beaches or dunes that remain stabilized for a very long time. Hence the production of this type of orthoquartzite requires stable tectonic conditions, characterized by very mild and gradual epeirogenic upwarps and downwarps to let the seas transgress and regress across the area. Under such stable conditions the source area is worn down to a peneplane; influx of detritus is slowed very markedly, so that winds, waves and currents are able to do their work thoroughly on each grain. Consequently these sediments are practically always supermature and show extreme roundness. Because detritus is contributed very slowly, and because no part of the depositional area is going down rapidly, the mild epeirogenic warpings and continual transgressions and regressions of the sea spread the deposits out into a thin blanket of a large area; sheets of sand 100-200 feet thick may cover 10,000 square miles. These sheets are made up of coalescing beaches and dune sands reworked into beaches or peritidal and shallow marine sands. The humid climate, given such a long time in which to work on the parent material in the deep soils of the source area, destroys all the feldspar. Hence this rock type consists entirely of quartz, almost all "common" quartz. If the source is deep within a batholith, there is very little vein quartz. Of course there is no chert and little metamorphic quartz. The only heavy minerals that can stand this prolonged abrasion are the ultra-stable tourmaline and zircon. The grain size is medium to fine sand (the typical size of beaches and dunes), and there is no silt or clay; gravel is rare, occurring only near the base. Quartz overgrowths are perhaps the most common cement, but carbonates are common and many of these rocks (even older Paleozoic ones) are uncemented. The color is very commonly white, but some are red due to hematite. Sometimes thousands of feet of highly mature to immature quartzarenite accumulate in narrow troughs bordering a stable Kraton. These sands, symbolically Q_k.G/g/MB_t may be littoral to turbidite sediments, and often form as the initial stage of KΛK, as a continental plate begins to split.

Plutonic Source, Arid Climate (or rarely, cold climate), frequently Q_k. G/s/Cd,L(n). These require much longer tectonic stability than the preceding type and are consequently rare. They also form thin blankets of supermature sand, formed in desert dune and beach environments (dune more important here) neighboring a peneplaned source land and are of course an extension of the climatic arkose. The distinguishing mark is the presence of extremely well rounded and frosted quartz and feldspar grains, the feldspar are smaller than the quartz and usually perfectly fresh (the "Dogma of the Immaculate Feldspar"). Little else but common quartz and well-rounded zircon and tourmaline comprise the sand. Cement may be authigenic quartz, feldspar overgrowths, carbonate, or evaporites. In other properties these rocks resemble the preceding type. There is likely to be little shale section associated with these rocks because of the lack of soil mantle in the source; the plutonic humid orthoquartzite frequently is accompanied by shale section (either stratigraphically adjacent or in a lateral facies development) because of the thick kaolinitic clay soils developed on intense weathering.

Metamorphic Source, often Q.M/s/L(w). These also develop under quiescent tectonic conditions and form widespread blanket sands, chiefly of beach-dune environments. Metamorphic materials are eliminated by prolonged abrasion, the only reminders being a high proportion of metamorphic quarts types and undulose quartz; hence these grade into the subphyllarenite. Pebbles are common, and consist of vein quartz and various types of metaquartzite. Sands are supermature to submature and of fine to medium grain. Immature quartzarenites form if beach sand is blown or carried into low-energy environments such as lagoons. No feldspar or chert is present. If the quartz is not rounded, heavies may contain some less stable metamorphic minerals like amphiboles, garnet or epidote, but often abrasion is long enough so that only rounded tourmaline and zircon remain. Associated shale section is rich in mica and micaceous paste, derived from breakdown of slate, schist, etc. Quartz or carbonate cement are common. The color is generally white or light gray, but is occasionally red.

Vein Quartz Source. Many areas of soft, low-rank metamorphic rock are shot through with abundant quartz veins. If such areas are eroded under quiescent tectonic conditions, the beach-dune activity is sufficient to wear away the soft MRF's so that about the only resistant particles left are the sand grains and pebbles composed of vein quartz, recognized by their richness in bubbles. This is an uncommon rock type, but there are a few conspicuous examples.

Older Sedimentary Source. Since no period of tectonic stability and no period of beach-dune action is required to attain a quartzarenite by reworking older, already quartz-rich sediments, these can form under any tectonic framework and almost any environment. Some are orogenic sediments, deposited in thick wedges and eroded from source areas consisting of older sandstones, mudrocks, and limestones; some develop on gentle uplift of a large area blanketed by older sedimentary rocks, hence form blanket sands. Maturity is generally low (because again, beach-dune action is not necessary for their production), and they are characterized primarily by many textural inversions (poor sorting and high rounding; or, more commonly, a lack of correlation between roundness and size, with mixture of angular and rounded grains within the same size, or small round grains plus large angular ones). The orogenic sediments are poorly sorted, having the size-distribution of a tectonic arkose, but the mineralogy of a quartzarenite to chert-arenite and include reworked rounded grains. Conglomerates are frequent, with pebbles of chert, sandstone, perhaps limestone, and inherited pebbles of vein

INTERPRETIVE (GENETIC): THE QUARTZARENITE

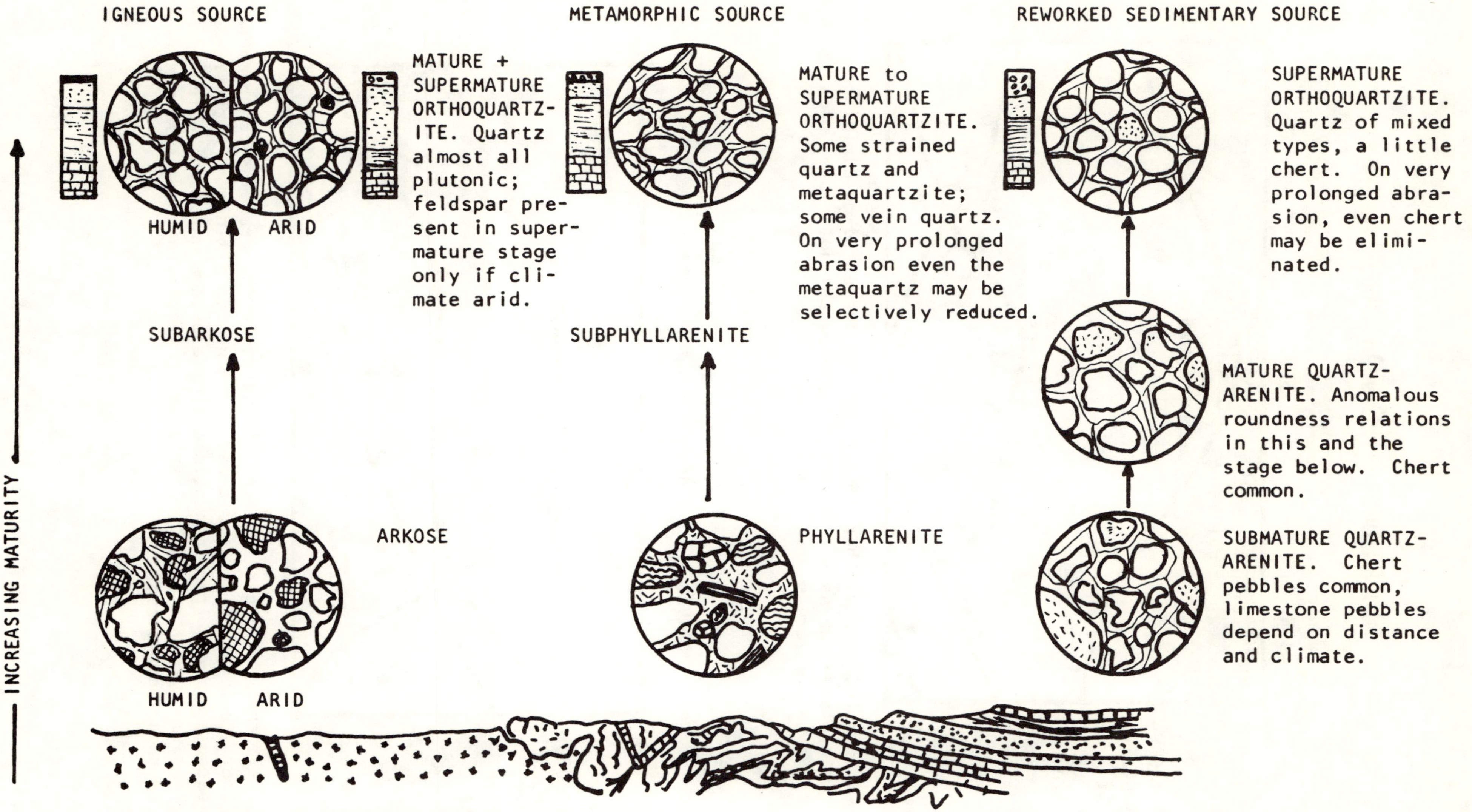

PROLONGED ABRASION OR WEATHERING REQUIRED to get an Orthoquartzite from these rocks which are low in quartz to start with.

ONLY BRIEF REWORKING NEEDED to get an orthoquartzite from these sediments which are already high in quartz.

MINERALOGY - THE QUARTZARENITE

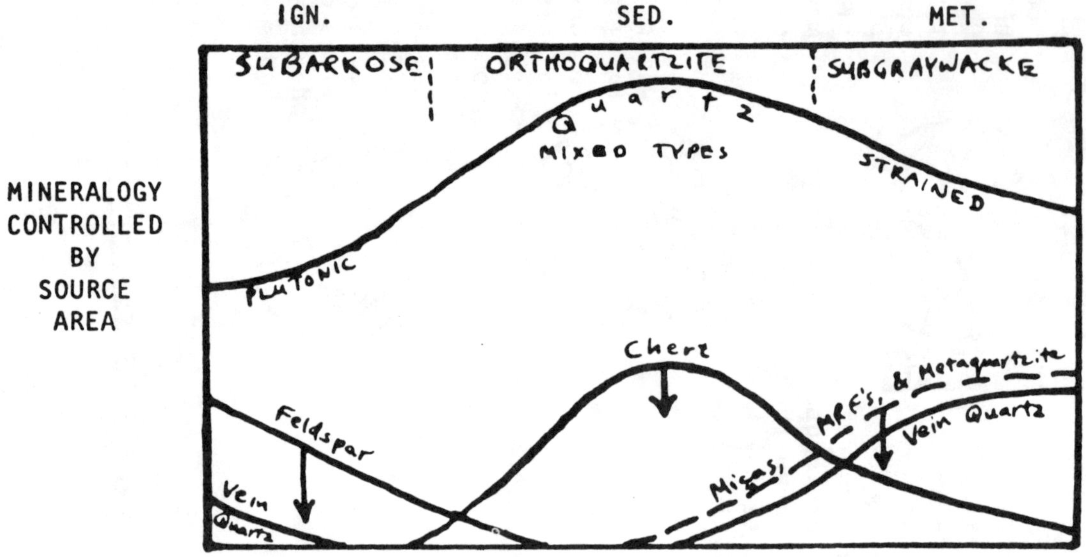

Arrows indicate whether constituent is increased or diminished on prolonged abrasion. Chert is usually but not always present in a sedimentary source area.

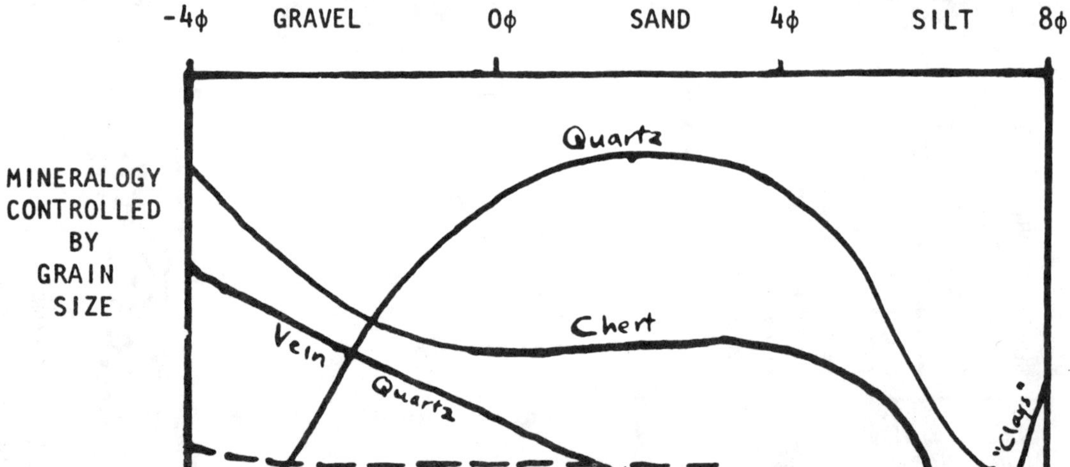

Chert and vein quartz are most abundant in the gravel sizes because they start out as large pieces. Vein quartz diminishes with size because some of its diagnostic criteria become unrecognizable in the fine-sand to silt range. Chert, being finer grained, is recognizable to much finer sizes.

quartz or metaquartzite. Because these are derived from erosion of many previous formations, quartz types are diverse and some reworked overgrowths may be found. In addition to inherited well-rounded grains, angular quartz may come from older, non-rounded sandstones. Abundant chert is the chief diagnostic material; it is quite commonly angular and associated with rounded and reworked quartz grains. There may be a very little feldspar, mica, etc., so that these are not quartzarenites of high purity, and grade into adjacent clans. The heavy minerals are also diverse but characteristically contain some recycled, rounded tourmaline and zircon. The color of these rocks is usually dirty white or light gray with dark specks (chert).

Petrology of Chert-Arenites

These rocks are members of the litharenite family, in which chert fragments are important. Some sandstones and conglomerates contain 50-90% chert grains, most of these are dark gray in hand specimens. They are usually orogenic sediments eroded from tectonically disturbed older sedimentary sequences, D.S/p/Cf. Mesozoic sandstones of the Rockies are often chert-arenites (Ballard). However, some chert-arenites are residual gravels and sands developed on extensively weathered cherty limestone terranes-Q.S/s/C(w).

Petrology of Calclithites, typically D. Sc/p/Cf(n).

Calclithites are terrigenous rocks of the Litharenite group, in which carbonate rock fragments dominate; often they contain over 50 percent carbonate fragments, obtained by erosion of limestones or dolomites outcropping in a source land. The source rocks are usually (though not necessarily) much older than the calclithite itself; thus the Miocene Oakville formation of Texas consists of grains of Cretaceous limestone; some Triassic beds in New Mexico consist of pebbles of Pennsylvanian Carbonate rock; Pennsylvanian conglomerates in Oklahoma consist of Ordovician limestone pebbles. Hence these are to be distinguished from intrasparite limestones, which are made up of a penecontemporaneous intraclasts torn up from the basin of depositon itself. Calclithites are terrigenous rocks like arkoses, graywackes, or orthoquartzites; it is indefensible to consider them as limestones although they have the composition of a limestone--it would be just as bad as calling an arkose a "granite" because they both contain quartz and feldspar.

In order for a rock to be made largely of a constituent so soft and so soluble as limestone, the rate of erosion must have greatly overbalanced the rate of chemical decay. Hence these rocks are exactly like arkoses, whose unstable and soft feldspar can be preserved only by rapid erosion or a dry (or cold) climate. Therefore calclithites also form chiefly in areas of intense faulting, but where the sedimentary cover is so thick or faulting is not of enough magnitude to get down to the granitic basement rock. It is possible that some calclithites are the result of a cold or arid climate alone, though these also require rather rapid erosion and deposition to preserve the fragments from much abrasion. Calclithites are common in the post-orogenic sandstones of thrusted cabonate regions such as the calcareous Alps and Carpathians of Slovakia and Poland (Siedlecka). In these areas of vigorous deformation, they have formed even under a humid climate, and are still accumulating today. Calclithites are forming today even in tropical New Guinea as a result of violent uplift (see Creaser, 1977, Austr. Atnl. Univ. for huge review).

The source area of calclithites must have had rugged relief, and the deposits are laid down in alluvial fans and river channels (like the tectonic arkoses). Contemporaneous faulting within a carbonate basin may produce marine calclithites and some calclithites are turbidites (McBride). Maturity is variable, and conglomerates are very frequent; in fact it appears that most calclithites are conglomerates. Consequently deposits accumulate as large prisms and decrease rapidly outward in thickness. Most calclithites (probably all) are cemented with calcite. They often contain rounded quartz grains, fragments of angular chert, sandstone, or shales, and of course are characterized by reworked fossils. Upon deeper erosion, basement rock may contribute feldspar or MRF's, and the rocks pass into arkoses or graywackes. Chert-arenites and calclithites form under similar tectonic conditions, and the difference between them may be partly due to climate and relief.

PETROLOGY OF MUDROCKS

Mudrock (word coined by Ingram) is a general term used herein to cover those terrigenous rocks that contain more than 50 percent silt and/or clay. The most obvious division of mudrocks is on the basis of texture and structure:

Grain size of mud fraction	Soft	Indurated, non-fissile	Indurated, fissile
over 2/3 silt	silt	siltstone	silt-shale
subequal silt and clay	mud	mudstone	mud-shale
over 2/3 clay	clay	claystone	clay-shale

(if the rocks contain over 10 percent sand, its median diameter should be prefixed to the above terms, e.g., very fine sandy siltstone, medium sandy clay-shale, etc. Soft yet thinly banded clays can be termed "laminated clay" where the term "clay-shale" would be undesirable because of the implied hardness. The writer has often used "silty clay-shale" or "silty claystone" for rocks containing 5 to 33 percent silt, reserving the unmodified term "claystone" for those with over 95 percent clay). See Picard 1971 J.S.P., for another scheme.

Mudrocks have long been ignored by petrographers, therefore not much is known about them. Partially this is due to a defeatist complex, caused by the very mistaken impression that they are too fine grained for anything useful to be seen with the microscope. When examined in thin section, however, they reveal a surprising diversity of texture, structure, and mineralogy.

Mineral Composition. If the rock contains more than a small percentage of silt, it can usually be related to one of the terrigenous rock clans. Orthoquartzite-type mudrocks have silt grains consisting almost entirely of quartz (or chert). Phyllarenite-type mudrocks are rich in fine-grained micas and may have some discrete metamorphic rock fragments and quartz silt showing composite or undulose extinction. Arkosic mudrocks contain much potash feldspar.

The nature of the clay mineral present often allows hints as to source-area lithology, climate, or environment of deposition (see section of this syllabus on clay mineralogy). The clay mineralogy should always be checked by X-Ray; the petrographic microscope can usually identify the dominant clay mineral, but those minerals present in small proportions will be missed. On the other hand, the microscope is superior in picking out clay minerals that are present in small patches but make up little in overall proportions of the rock (e.g. authigenic chlorite or kaolinite patches).

Well-sorted (mature) siltstones of course have the same cements as the corresponding sandstones: carbonates, quartz, iron oxides, gypsum, etc. But thin sections of clay shales often show pores and irregular openings filled with authigenic quartz, carbonates, barite, gypsum, and other minerals; some of these have grown within the yielding clay by physical displacement of the mud.

Mudrocks contain many miscellaneous transported constituents. Fossils including non-calcareous objects such as spores or spicules, are not infrequent. Glauconite, Phosphatic pellets, etc., are sometime present.

Structures. Shales have an astonishing variety of structures. Besides such "normal" structures as thin lamination, graded bedding, slumped bedding, small-scale cross-bedding, cut-and-fill structures (mostly invisible in hand specimen), shales contain many peculiar ones. Many are revealed in thin section to be really micro-conglomerates, made up of aggregates of soft clay-balls or curdled, vague lumps. Others show apparent desiccation features and irregular shrinkage (?) cracks. Balls of concentrically-layered silt, slumped, curdled, or otherwise irregular stringers and isolated lumps of sand are common in some shales. Burrows of animals riddle many specimens of shale, and some contain clay fecal pellets.

In most clays, the clay minerals are oriented so that the flakes lie parallel with the bedding. Orientation occurs in clays ranging in age from Cambrian to Pleistocene; in depth of burial from tens of thousands of feet to a few tens of feet. Therefore it is not a diagenetic feature that requires great pressures or geological age, rather it forms upon original deposition of the clay. Some claystones have randomly oriented clay minerasl; these also may be of any age or depth of burial. The factors that cause orientation are not now known, but organic burrowing is probably the chief cause of unoriented clays.

For a claystone to be fissile requires that all the following conditions be satisfied: (1) the rock must contain little silt or sand; (2) the rock must show even bedding undisturbed by slumping or burrowing organisms; (3) it must contain little or no chemical cement; (4) the clays must be well oriented. If any one of these conditions is not satisfied, the clay will not be fissile. Yet sometimes one obtains a well-bedded, "pure" claystone, lacking in cement and with oriented clays that still is not fissile despite the fact that it fulfills all above requirements. The presence of mica or very thin laminae of fine silt is an aid to fissility.

DESCRIPTION AND NOMENCLATURE FOR TERRIGENOUS SEDIMENTS

For super-detailed description of sedimentary rocks of all types in thin section, the following outline is used. This form sheet covers in complete detail every rock property, and for many rocks, parts of it may be omitted.

Before starting with the microscope, hold the slide up and look at it with a hand lens to see the major structures. Then start with the lowest microscopic power. It is very useful to look at slides using direct sunlight as a reflected light source; much valuable detail can often be seen in otherwise featureless opaques like limonite or hematite.

Point-counting technique and theory are discussed by Chayes (1956). A point count is most conveniently accomplished by laying out an equispaced grid to cover most of the microscopic slide, but theoretically any random distribution of points is satisfactory, just so one does not cheat in the location of points. All points that the crosshair lands on should be tallied, even "unknowns" or "garbage". The interval between points should be larger than the largest grain; but if this cannot be managed and one grain is hit twice, it should be counted twice, etc. For most sandstones a spacing of about 1 mm is satisfactory. In a thin section, the traverse should not run parallel to bedding because a string of points might run along a placer of micas, heavy minerals, etc. For coarse rocks, a grid of lines can be ruled on plastic film and placed over the sawn surface.

I. REFERENCE NUMBER, Geologic age, formation, stratigraphic level within that formation, locality, regional geology and structure (for example, if the rock comes from a very tight fold or from vicinity of a fault that will affect its properties and should be noted).

II. NAME OF THE ROCK (following the pattern, grain size: prominent orthochemical cements, textural maturity, notable or unusual transported constituents, main rock name. This of course is done last).

III. MEGASCOPIC PROPERTIES (hardness, color, freshness, sedimentary structures, fossils, sorting and grain size, composition of the terrigenous and orthochemical minerals).

IV. MICROSCOPIC DESCRIPTION
   A. Brief summary of the important features of the rock (so the reader doesn't have to wade through a great mass of detail to find the features you think are very important--give a brief summary of the major, unusual or interesting features of the rock just like you'd write an abstract of a paper).

   B. Texture
      1. Fundamental end-members
         a. Percent terrigenous materials
         b. Percent allochemical materials
         c. Percent orthochemical materials
         d. Main rock group, based on the above (See page 2, syllabus).

2. Fabric
   a. General homogeneity (Can the entire specimen be considered a single rock type, or are two or more types interlayered--such as sandy streaks in a mudstone, or bioclastic layers in an otherwise non-fossiliferous sandstone?)
   b. Packing (Is there any evidence of welding or interpenetration of clastic grains by solution on the one hand, or spreading of grains by the growth of cements on the other hand?)
   c. Porosity both before and after cementation
   d. Perfection of orientation and how it is expressed (i.e. by micas only, or by micas and elongated quartz grains, etc.)
3. Grain size
   a. Entire sediment: give median, extreme (100%) range, 16-84% range, and sorting in terms of $\sigma\,\phi(\phi\,84-\phi 16)\,/2$. Sizes should be given first in mm (if you like they can be converted to ϕ also). Is the distribution unimodal or bimodal, and give details as to the diameter of the modes and their relative proportions in the sediment. If there is any correlation of size with composition, describe also.
   b. Gravel fraction: give percent, median, sorting, range.
   c. Sand fraction: give percent, median, sorting range.
   d. Mud fraction: give percent, relative proportion of silt versus clay, and median of silt if determinable.
   e. Complete textural name (use the triangular diagram; the final name should be e.g. pebbly claystone, or silty medium sandstone, etc.)
4. Grain shape
   a. Idiomorphism, range of idiomorphism, and variation of this property with composition.
   b. Sphericity (or elongation), sphericity range, and variation of sphericity with size and composition.
   c. Roundness, roundness sorting, and variation of roundness with size and composition.
5. Textural Maturity. Define the stage of textural maturity; are any inversions present (such as well-rounded but poorly-sorted grains)? Immature stage: the rock contains over 5% clays and very fine micas under 0.03 mm. Submature stage: Clays less than 5%, but the rest of the rock is still poorly sorted--i.e., the 16-84% grain size range is more than 1.0 phi units or Wentworth grades. Mature stage: rock is well-sorted (16-84% range less than 1.0ϕ) but still not well-rounded. Supermature stage: rock is well-sorted and the quartz grains of sand size show an arithmetic mean roundnes of 3.0ρ or better.
6. Bonding agents. Relative effectiveness of each cement (or the clayey matrix) in bonding the rock together. Cements are described in detail later, under C3.

C. Mineral Composition

Use the following detailed outline for describing each mineral present. It is a great aid to describe the minerals in systematic order every time. The following order of description is suggested:

1. Terrigenous Minerals
   a. Quartz (stretched metaquartzite should be counted separately and its percentage specified because it is sometimes assigned to the "M" or "RF" poles, not the "Q" pole).
   b. Chert
   c. Feldspar
   d. Mica
   e. Rock fragments (metamorphic, plutonic or volcanic, reworked sedimentary). Their internal composition texture and structure should be described, and they should be specifically identified if possible (e.g. muscovite schist, andesite, biomicrite, illite-slate).
   f. Other terrigenous minerals--such as tourmaline, zircon, etc.; each of these should be described separately.
   g. Clay minerals.
   At the end state what percentage you allot to the Q, F, and R poles of the composition triangle, and give the compositional rock name (e.g. arkose, or feldspathic phyllarenite, etc.). This should be based on a point-count of at least 100 essential (QFR) grains.
2. Allochemical Grains (here describe separately the fossils, oolites, intraclastic particles, pellets, collophane, glauconite and other allochems, using the complete outline 1 - 25 below).
3. Orthochemical minerals. Also describe each variety of these individually, using the following outline, 1 - 25.
   For each of the properties, for each of the minerals, ask yourself <u>why</u> this is so; for example, if zircon is smaller than quartz and more angular, and tends to be placered into streaks, you reason that it is smaller and placered because it has a higher density; also it is more angular than quartz because it is harder. A great amount of valuable information can be unearthed if you will continually ask yourself <u>why</u>; it will stimulate you to look for clues that otherwise might be completely missed. This is the outline to be followed for each mineral (Terrigenous, Allochemical, and Orthochemical) in the order set forth above.
   1. Name. If the mineral varies in composition (such as plagioclase or hornblende) identify the composition optically.
   2. Method of identification if the identity is not obvious.
   3. Percentage present in section. State method, whether sheer estimation, point-count, etc. A ± error (σ) should always be attached.
   4. Occurrence in slide and distribution pattern (uniformly distributed, randomly scattered, or patchily distributed. For example, heavy minerals often show a patchy distribution as they are concentrated along certain bedding planes as placers; or certain cements may occur in patches and be absent in the rest of the rock; or certain minerals may occur only in the vicinity of joints or near weathered surfaces; or certain minerals like pyrite may be selectively associated with the more clayey layers, etc.).
   5. Physical orientation
   6. Grain size of this mineral: median, extreme range, and sorting; if desirable the 16-84% range should be specified.

Is the size distribution continuous (unimodal) or discontinuous (bimodal)?

7. Idiomorphism, and range of idiomorphism. If the mineral is nearly idiomorphic, describe the crystal habit if desirable. Are grains truly crystals or merely cleaved pieces?
8. Average Sphericity (stated as W/L, e.g. .67), uniformity and range of sphericity.
9. Average roundness, (Powers classes), uniformity and range of roundness, and variation of roundness with grain size. Is there a continuous gradation in roundness, or are there rounded and angular grains with no intermediate roundness values?
10. Etching, surface and contact features; is the mineral indented by others, or does it indent its neighbors? (This is important in paragenesis).
11. For clastic minerals, are overgrowths present? For authigenic minerals, are clastic nuclei present?
12. Cleavage, fracture, or parting.
13. Zoning and internal structure (describe in detail).
14. Index and relief
15. Color, pleochroism, diaphaneity; luster and color in reflected light.
16. Twinning.
17. Birefringence; does it show undulose or straight extinction?
18. Optical elongation and orientation; extinction angle.
19. Interference figure: uniaxial or biaxial, plus or minus, 2V, dispersion.
20. Inclusions
    a. Identity of each inclusion
    b. Properties of each, using above outline for description
    c. Size of each inclusion
    d. Distribution or zoning of inclusion within the host grain, and orientation of inclusion within host
    e. Abundance of each inclusion type
21. Freshness, alteration products, homogeneity of decomposition with the same species, and time of alteration (predepositional, epigenetic etc.--give the evidence).
22. Definition and description of varieties within one mineral species; differences in other properties that may occur between varieties (for example, is quartz with rutile needles more round than quartz without rutile needles; is purple zircon larger than colored zircon; etc. This is very important in determining history of rock). What is the relative abundance of each variety?
23. Affinities and antipathies of occurrence (e.g. does calcite selectively occur in the coarser-grained layers; is dolomite preferentially associated with organic matter, etc.).
24. Chemical and age relations with other minerals (i.e. is it being replaced or attacked by other minerals, is it euhedral against others, earlier or later than others, at what time and how did it form; give evidence for these statements).
25. Derivation and probable source area for detrital constituents.

D. Structures, etc.

1. Sedimentary structures (bedding, lamination, cross-bedding, slumping, etc.).
2. Tectonic structures (joints, etc.).
3. Weathering and alteration.

E. Interpretation and Paragenesis

To a large extent these conclusions will be tentative until a rather large suite of related samples has been built up. Nevertheless it is possible to gain a great amount of information from one or a few slides. To that aim the following outline is suggested:

1. Source area
   a. Geology (based upon quartz types, presence and abundance and character of all the minerals present). Give estimate of the proportions of the rock contributed by each source.
   b. Relief and tectonic state (based upon mineralogy, grain size, maturity).
   c. Climate (based chiefly on feldspars and micas and heavy minerals and clay minerals).
   d. Length of transport or distance of source area (based partly upon size of largest grains, and shape).
2. Depositional area
   a. Environment of Deposition (beach, river, delta, dune, lagoon, etc.)
   b. Depth of water, strength and persistance of currents, salinity, rapidity of burial, effect of organisms).
3. Diagenetic and post-diagenetic changes
   a. Age relations and mode of origin of authigenic constituents.
   b. Effects of intrastratal migrating fluids.
   c. Effects of post-emergent weathering.

F. Economic Importance

1. What are the potentialities of this rock as a reservoir for oil or water? What is the present porosity and how did it originate? What is the grain-size of the pores and how well are they interconnected? Does the petrographic character (e.g. whether phyllarenite or orthoquartzite, or type of cementing material) suggest a favorable porosity trend in a given geographic direction?
2. What valuable clues does the specimen offer to correlation in the way of odd minerals or characteristic varieties of minerals?

For rapid descriptions, skimping should be done only under mineral description; but one should notice especially the following items about these minerals: Quartz (possible bimodality in size; average roundness and range; is there a break in the roundness distribution; correlation of roundness with size; dominant extinction and inclusion types; what is the relative amount of plutonic versus metamorphic and vein volcanic types). Feldspar (size and shape with respect to quartz; average freshness and range of freshness within the same species--look especially for kaolinization; do the several species of feldspar present differ as to average freshness, size, or roundness). Chert (size and shape with respect to quartz; are several varieties present, such as oolitic, spiculitic, etc.). Micas (abundance of the several

NOMOGRAPH FOR DETERMINATION OF COUNTING ERROR

Adapted from data of Van der Plas & Tobi, (1965) AmJSci 263:87.

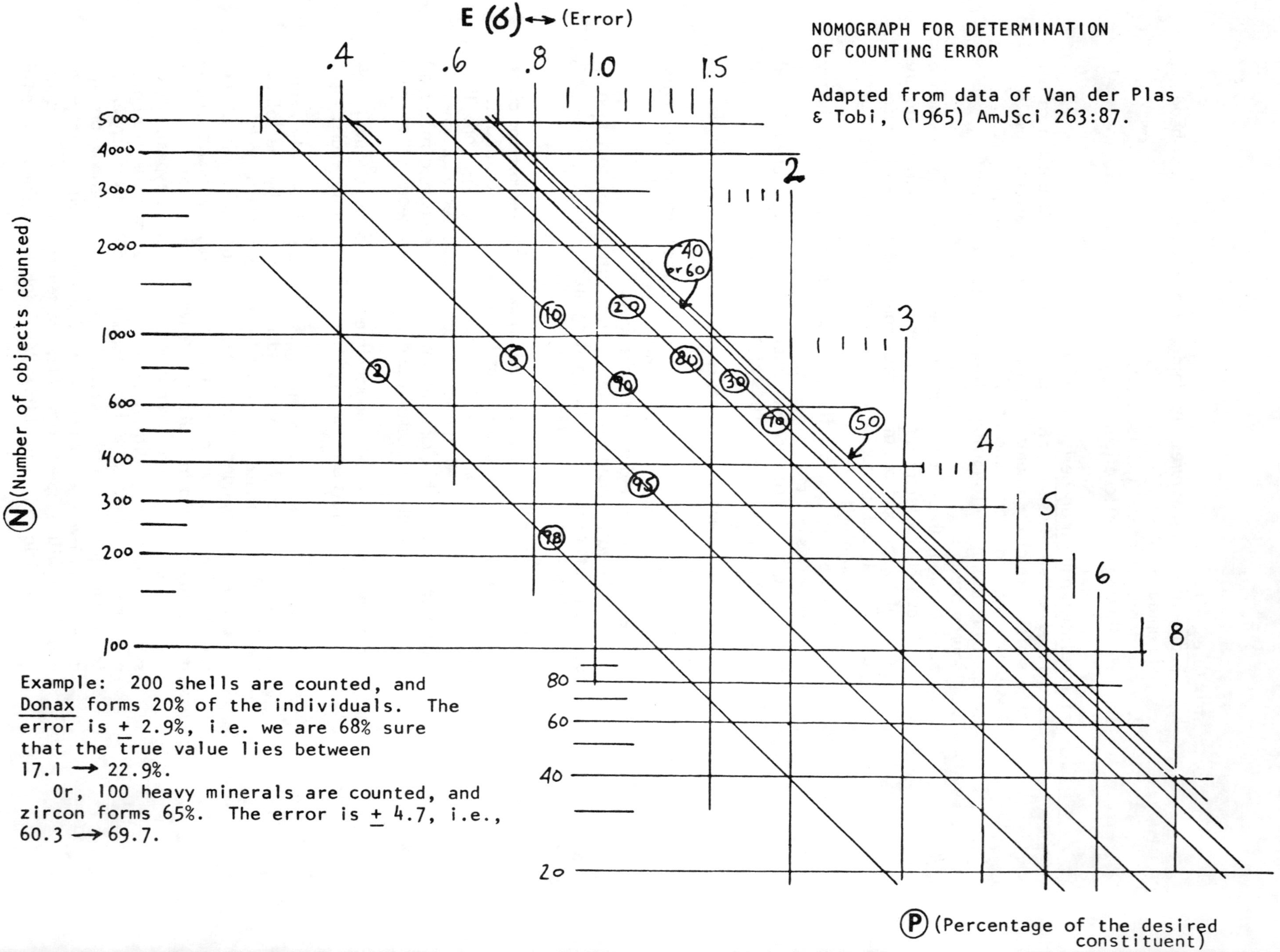

Example: 200 shells are counted, and Donax forms 20% of the individuals. The error is ± 2.9%, i.e. we are 68% sure that the true value lies between 17.1 → 22.9%.

Or, 100 heavy minerals are counted, and zircon forms 65%. The error is ± 4.7, i.e., 60.3 → 69.7.

types). Metamorphic rock fragments: (rank; size and shape with respect to quartz). Authigenic minerals (time and mode of origin).

Petrographic Nomenclature for Conglomerates, Sandstones, and Siltstones. Progress in a science can often be measured by the increase in precision of its nomenclature. Much more information is communicated if we say "Spirifer mucronatus" rather than "brachiopod" or simply "fossil" as was done in the 17th century. Describing an igneous rock as "phanerite" or "trap rock" long ago passed out of style, and we now use such terms as "pink, fine-grained hornblende quartz monzonite." Unfortunately, primitive and simple (thereby meaningless) terminology still prevails among geologists for sedimentary rocks, and most stratigraphic sections are still described simply as sandstone, shale, or limestone. To do this is to gloss over the tremendous variation which occurs in these rocks, and to obscure stratigraphic and regional changes, most of which would become of obvious importance if proper terminology were in use. How far would geologic work in igneous and metamorphic areas progress if areas and sections were described as covered with outcrops of "fine, pink rock," "course rock," or "crystalline rock" as is now done with sediments?

The more information a term communicates, the longer it must become. A term including all the important information about a sandstone, then, is bound to be lengthy because there are so many important ways in which sandstones can vary. Five sandstone properties are held to be important enough to be included in the rock name: the grain-size name, the chemical cements, the textural maturity, the miscellaneous transported constituents, and the clan designation.

1. The grain-size name is the most obvious property to many geologists, and often is the only description given to a sandstone. First determine the percentage of gravel (grains larger than 2 mm) if posible from a hand specimen because the thin section is too small in area to give a representative sample of the gravel content. Then estimate the median size of the gravel fraction alone. From this point on, the process is the same whether the rock contains any gravel or not. Estimate the ratio of sand to mud (silt plus clay). This can be easily done by selecting a detrital grain .062 mm in diameter and placing it under the cross-hairs to use as a standard for comparison, then estimating what proportion of the rock consists of grains larger than this. Next, estimate the median size of the sand fraction, alone. Finally, estimate the ratio of silt vs. clay matrix and name according to the triangle on page 28.

2. The chemically-precipitated cements offer important clues to the chemical conditions prevailing when the rock was deposited or afterward. Clay is not considered as a cement, but as a detrital matrix. If one considers clay as a cement, he gets into difficulties with the shales which then become entirely "cement." Only the prominent cements (usually one, almost never more than two) are included in the five-fold name. Wherever there is a possibility of confusion in that one mineral can be present in a sandstone either as a cement or as a transported constituent, use the construction "-cemented" as opposed to "-bearing". The following usage is suggested:

Quartz: If the rock is so tightly cemented with abundant quartz overgrowths that no pore space is left (and the rock consequently breaks through the grains), use "highly siliceous." If there are only a few small overgrowths, use "slightly siliceous." Most rocks may be called simply "siliceous."

Chert: "Chert-cemented" (not "cherty" because this could mean a chert-bearing sandstone with chert sand grains). If the cement is well-developed fibrous chalcedonic quartz, then "chalcedonic" should be used instead.

Opal: "Opaline."

Carbonates: "Calcitic," "dolomitic," "sideritic", etc.

Iron oxides: If iron oxides from a mere superficial film or if they occur as a stain in a red clay matrix, they are not volumetrically important enough to include as part of the name. Only if the minerals form thick, pure colloform crusts or pore fillings should the adjectives "hematitic" or "limonitic" (if in doubt, "ferruginous") be used.

Clay minerals: if the clay minerals are detrital or have formed by recrystallization of a previous clay matrix, they are not considered as a cement, and their presence is denoted in the maturity term. Only if they are chemical precipitates, filling previous pore space (usually in the form of accordion-like stacks or fringing radial crusts) should they be included as "kaolin-cemented," "chlorite-cemented," etc. Usually even these are present only in small quantities and will be ignored in naming the rock.

Miscellaneous minerals: As the need arises, such terms as "pyritic," "collophane-cemented," "glauconite-cemented," "gypsiferous," "anhydrite-cemented," "baritic," "feldspar-cemented" etc. may be used.

If there is no cement, eliminate that part of the name.

3. Textural Maturity reflects the physical characteristics and has already been discussed at the beginning of the petrology section. The maturity term is placed third in the rock name, because it then conveniently separates two mineral terms.

4. Miscellaneous transported constituents are often of value in interpreting source, environment, and often are an important descriptive feature. No mineral whose presence is already implied in the clan name is included, i.e. quartz, feldspar, and metamorphic rock fragments are never used as modifiers. It is a matter of individual judgment when an accessory mineral becomes of enough importance to include as part of a fivefold name; in many rocks this part of the name may be omitted. The limits below are reasonable suggestions.

"Chert-bearing": Because chert is a very important mineral whose presence indicates a sedimentary source area in part, this adjective should be used if the rock contains more than about 5 percent chert.

"Volcanite-bearing": If 2-5 percent of volcanic rock fragments or volcanic glass is present, then this adjective is used.

"Micaceous": This adjective is used if discrete flakes of coarse mica are conspicuous (say 1-2 percent). If desired, "muscovitic," "biotitic" or "chloritic" may be used if these are dominant.

"Glauconitic": "Collophane-bearing," "Fossiliferous," etc., are used if they constitute more than 1-2 percent of the rock.

"Calcarenitic" or "calciruditic" are used if they form more than 1-2 percent of the rock. The adjective "calclithic" has been suggested for these (Anan). These terms are used to describe reworked fragments of older limestones, eroded from terrigenous sources. If they form penecontemporaneous limestone deposits, the terms "intraclast-bearing," "pelletiferous," or "oolitic" may be used. If over half the transported constitutents are reworked carbonate rock fragments, then the rocks are a member of the calclithite clan.

5. The Clan designation (arkose, quartzose phyllarenite, etc. analogous to granite or quartz diorite in igneous rocks) is the most important as it indicates the bulk mineral composition. Mineralogy depends chiefly on source area lithology, a factor which stays relatively constant for long periods of time and consequently is the most characteristic feature of a sandstone format ion. It is a more dependable property than the other parts of the name, which tend to fluctuate more rapidly. The clan name is determined as shown previously by using the composition triangles.

Order of terms

No order is inherently more sacred than any other; however, it facilitates comparison of samples if the same order is followed by everyone. As in igneous or metamorphic rocks the most important part of the name, the clan designation, should come last. Grain size, the one sedimentary rock characteristic that is used by all geologist, should come first. By convention, the following order is suggested for the fivefold name (asterisks denote sections of the name that can be omitted if the constituents are absent):

(Grain Size): (Cements*) (Textural Maturity) (Miscellaneous Transported Constituents*) (Clan Designation). Following are some examples:

Fine sandstone: siliceous submature glauconitic quartzarenite.

Pebbly medium sandstone: hematitic bimodal supermature chert-bearing subphyllarenite.

Silty very fine sandstone: gypsiferous mature arkose.

Sandy granule conglomerate: calcitic submature calclithite.

Muddy fine sandstone: immature phyllarenite.

Slightly granular medium sandstone: calcitic, siliceous mature micaceous phyllarenite.

Clayey very fine sandstone: immature fossiliferous plagioclase arkose.

Fine sandstone: supermature quartzarenite.

Silty coarse sandstone: chert-cemented submature quartzose phyllarenite.

Very fine sandstone: opaline mature chert-bearing and calcarenitic quartzarenite.

Mudrocks

For siltstones and silt-shales, the same terminology is used as above (e.g. siltstone: siliceous submature glauconitic phyllarenite). For mudstones, the order is inverted with the term mudstone coming last (e.g. immature chert-bearing subarkose mudstone, or calcitic immature orthoquartzite mudstone). Both of the above can usually be easily related to one of the terrigenous rock clans. For claystones, omit the textural maturity and clan names and follow the pattern (chemical cements) (odd clastic constituents) (chief clay mineral or minerals) (grain size). For example, calcitic fossiliferous illite-chlorite clay shale; chert-cemented collophane-bearing illite claystone. A classification has also been proposed by Picard (1971 JSP).

PETROLOGY OF CARBONATE ROCKS

Carbonate rocks comprise some 25-35 percent of the stratigraphic section. They are quite complex because of the varied constituents that form them and the amount of replacement and recrystallization that they undergo. Books by Bathurst (1976), Milliman (1974), Wilson (1975), and Scholle (1978) are excellent references.

Carbonate rocks may be best studied by the following means: (1) etching, wherein the specimen is immersed in dilute hydrochloric acid for say five minutes; this reveals distribution of relatively insoluble constituents such as sand, silt, clay, glauconite, phosphates, authigenic quartz, feldspar, chert, pyrite, and dolomite; and also reveals the morphology of the calcite, whether ooze, spar, fossils, intraclasts, etc. (2) Peels, wherein an acetate impression is made of the slightly etched surface; this can then be examined under a petrographic microscope in transmitted light and a wealth of textural detail is brought out in the calcite, in some ways superior to a thin section. (3) Insoluble residue, wherein the rock is dissolved and the insoluble constituents described under binocular and petrographic microscope. (4) Thin section, the best all-round method and absolutely necessary. All methods should be used to study a specimen adequately. Electron microscopy is also very useful (Longman & Mench 1978 Sed. Geol.).

The following scheme for carbonate classification is presented in more detail in January, 1959, A.A.P.G. Bulletin, and A.A.P.G. Memoir #1 (1962).

Classification of Limestones

Constituents. Disregarding admixture of terrigenous sand, silt, and clay, there are three basic components of limestone: (1) microcrystalline calcite ooze, (2) sparry calcite, usually a cement, and (3) allochems. These are, in most limestones, analogous with the clay matrix, chemical cement and sand grains of a sandstone.

(1)Microcrystalline ooze, forming grains 1-4 microns in diameter, usually sub-translucent to almost opaque in thin sections. In modern sediments, it may consist of calcite, magnesian-calcite, or aragonite; in pre-Pleistocene limestones, it is almost always converted to calcite. Carbonate mud is polygenetic. Some is the result of inorganic chemical precipitation in warm, shallow, saline water (probably the situation in the Persian Gulf). Most mud today comes from disintegration of organisms, mainly green algae, into constituent crystallites. Some mud also is windblown dust, can come from abrasion of shells, or may be precipitated by algae, bacteria, etc. The origin can sometimes be deciphered by study with electron microscope, isotopes, etc. Carbonate mud today forms in broad, warm shallow protected shelves, bays or peritidal flats such as in the Bahamas, Florida Bay, Trucial Oman, etc. Lithified ooze, or micrite, is the main constituent of the "lithographic" limestones.

(2) Sparry calcite forms crystals generally over 10 microns in diameter, and is distinguished from micrite by its clarity as well as coarser crystal size, which may range up to 1 mm or more. This type of calcite usually occurs as a pore-filling cement; pores are often lined first by fibrous calcite, with equant mosaic calcite filling the final spaces. In modern sediments, the first fibrous coating is usually aragonite or magnesium-calcite, but in older limestones all cement is converted to calcite. Some sparry calcite is the result of recrystallization of micrite; this subject is not touched upon here. For criteria see Folk (1965 SEPM Spec. Publ. 13), and Bathurst (1971).

Allochemical constituents (allos: differentiation from the normal) are those components that have formed by chemical precipitation within the basin of deposition, but which for the most part have suffered some later transport; or, if they have not been transported, they include such organized aggregates as sedentary fossils or fecal pellets differentiated from "normal" chemical precipitates as one usually thinks of them. Only four types of allochems are of importance: intraclasts, oolites, fossils, and pellets.

(1) Intraclasts represent pieces of penecontemporaneous, usually weakly consolidated carbonate sediment that have been torn up and redeposited by currents (hence the term intraclast, signifying that they have been broken from within the formation). Intraclasts vary from fine sand size up to pebble or boulder size, as in the familiar limestone conglomerates. Intraclasts themselves may consist of any kind of limestone, micrite, biomicrite, intrasparite, pelsparite, etc. (see later). They indicate a tearing-up of the bottom by an increase in current velocity (such as in storms), lowering of wave base by partial emergence, or possible tectonic instability of the basin of deposition. Rare intraclasts form by accretion, analogous to lumps in a sugar bowl, and are then transported. Of such nature are the "grapestone" aggregates of fecal (?) pellets in the Bahamas (Illing). They may also form in the wave-attacked margins of broad areas where ooze is being precipitated, or upon tidal and supratidal flats.

(2) Oolites range from 0.1 - 1.0 mm in diameter, and show radial or concentric structure. They usually form around a nucleus, often a small fossil fragment, fecal pellet, or quartz grain. They form in loci of vigorous and continuous current or wave action, in places where aragonite is being rather rapidly precipitated. An individual oolite may go through several cycles of abrasion alternating with growth. In the Bahamas (Illing) oolites form in current-swept tidal channels or form submarine dune ridges near shore where currents are powerful. Modern oolites are commonly aragonite, rarely Mg-calcite. Ancient ones may have been calcite (Sandberg).

(3) Fossils are important constituents of many limestones. The most common rock-formers are algae, foraminifera, sponge spicules, corals, bryozoans, brachiopods, gastropods, pelecypods, ostracods, trilobites, echinoids. See Horowitz and Potter for details.

(4) Pellets are homogeneous aggregates of microcrystalline calcite, well rounded and sorted, averaging .03 to .20 mm. They probably represent fecal pellets of worms or other invertebrates. It is possible that some may form in place by a form of recrystallization; vaguely defined "pellets" in ooze are sometimes termed grumeleuse.

Rock Classification. Almost all carbonate rocks contain more than one type of material; one may be a mixture of oolites, fossils and sparry calcite cement while another may consist of quartz silt, pellets, and microcrystalline ooze partially replaced by dolomite and chert. Thus the problem of classification becomes one in systematizing these variations of composition and drawing significant limits between types. Carbonate rocks are so complex that it is usually necessary to make a thin section study in order to pigeon-hole a specimen properly, although good estimates can be made in the field or with an etched specimen.

Disregarding for a moment the content of terrigenous material, it is possible to base a practical limestone classification on the relative proportions of three end-members (1) allochems, (2) microcrystalline ooze, and (3) sparry calcite cement.

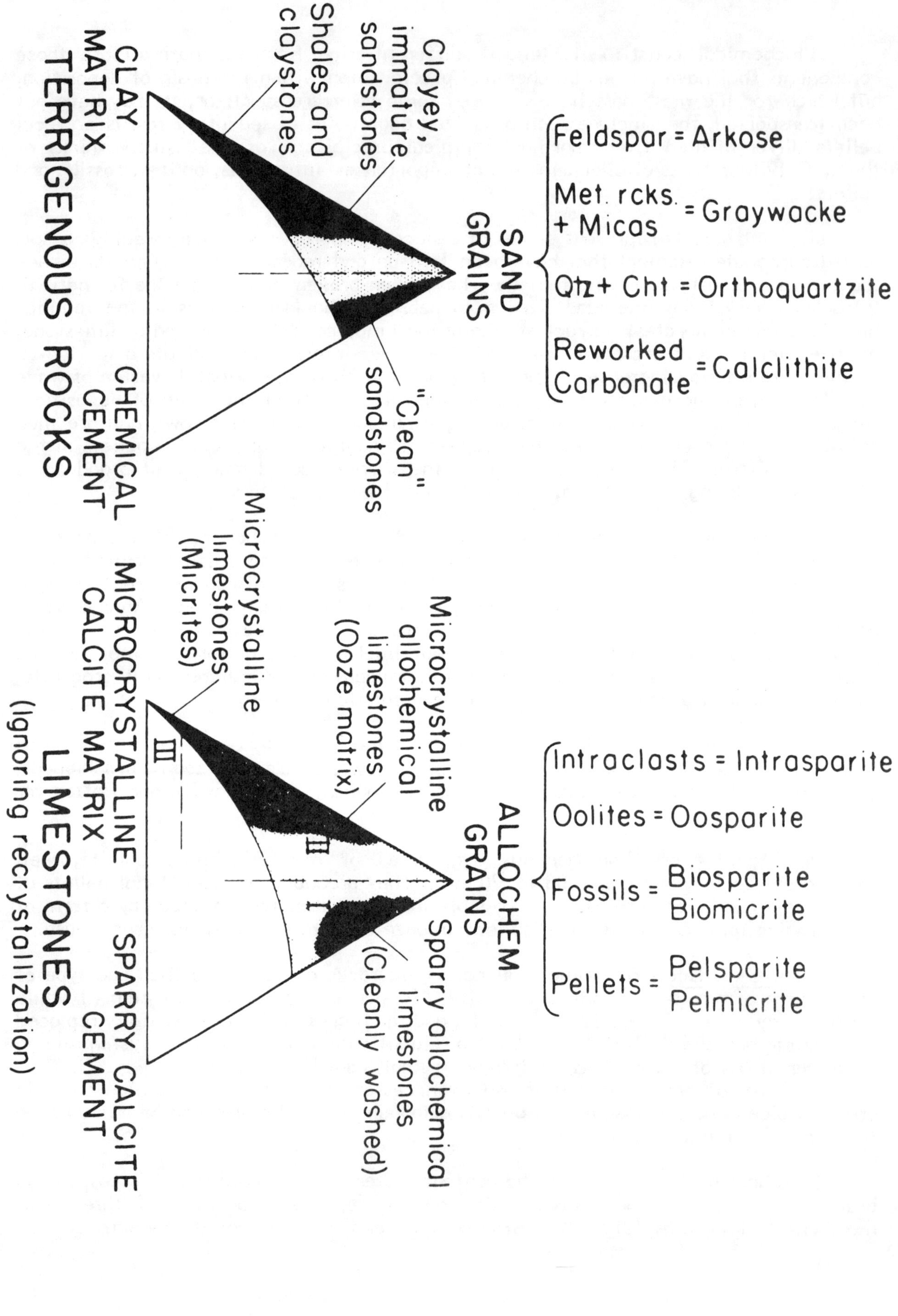

SAND GRAINS
Feldspar = Arkose
Met. rcks. + Micas = Graywacke
Qtz + Cht = Orthoquartzite
Reworked Carbonate = Calclithite
Clayey, immature sandstones
Shales and claystones
"Clean" sandstones
CLAY MATRIX
CHEMICAL CEMENT
TERRIGENOUS ROCKS
ALLOCHEM GRAINS
Intraclasts = Intrasparite
Oolites = Oosparite
Fossils = Biosparite Biomicrite
Pellets = Pelsparite Pelmicrite
Microcrystalline allochemical limestones (Ooze matrix)
Microcrystalline limestones (Micrites)
Sparry allochemical limestones (Cleanly washed)
I
II
III
MICROCRYSTALLINE CALCITE MATRIX
SPARRY CALCITE CEMENT
LIMESTONES
(Ignoring recrystallization)

Allochems represent the framework of the rock: the shells, oolites, carbonate pebbles or pellets that make up the bulk of most limestones, analogous to the quartz sand of a sandstone or the pebbles of a conglomerate. Microcrystalline ooze represents a clay-size "matrix" whose presence signifies lack of vigorous currents, just as the presence of a clay mineral matrix in a sandstone indicates poor washing. Sparry calcite cement simply fills up pore spaces in the rock where microcrystalline ooze has been washed out, just as porous, non-clayey sandstones frequently become cemented with chemical precipitates. Thus the relative proportions of microcrystalline ooze and sparry calcite cement are an important feature of the rock, inasmuch as they show the degree of "sorting" or current strength of the environment, analogous to textural maturity in sandstones. If we plot these two constituents and the allochemical "framework" as three poles of a triangular diagram (see figure), the field in which limestones occur is shown by the shaded area; divisions between the three major textural types of limestone are also shown on this figure. A similar field appears if one plots terrigenous rocks on a triangle with the three poles of sand and silt, clay and orthochemical cement.

This classification is predicated on the assumption that the sparry calcite has not been formed by aggrading recrystallization of a fine calcite ooze, and that microcrystalline calcite has not formed by degrading recrystallization of coarser calcite. In most carbonates the writer has examined, this assumption is believed to be very largely true. Nevertheless, the writer agrees that recrystallization is a very important process in some limestone formations, and the classification proposed here does not apply to recrystallized rocks. It provides a starting point for study of recrystallized rocks, though, because on original deposition these rocks must have belonged to one of the groups here proposed.

Type I limestones (designated as Sparry Allochemical rocks) consist chiefly of allochemical constituents cemented by sparry calcite cement. These rocks are equivalent to the well-sorted terrigenous conglomerates or sandstones in that solid particles (here intraclasts, oolites, fossils or pellets) have been heaped together by currents powerful or presistent enough to winnow away any microcrystalline ooze that otherwise might have accumulated as a matrix, and the interstitial pores have been filled by directly precipitated sparry calcite cement. The relative proportions of sparry calcite cement and allochems varies within rather restricted limits because of the limitations of packing:

(1) There is a limit to the tightness with which allochems may be packed, thus there will always be some pore space available for cement to fill, and

(2) There must be a certain minimum amount of allochems present in order to support the structure-sparry calcite cement grows only in poor spaces and in general cannot form a rock on its own right, unless recrystallization occurs. Similarly sandstones require a minimum amount of sand grains (say 60 percent) to support the rock structure. It may be noted that carbonate rocks on deposition often have porosity much greater than sandstones or conglomerates of equivalent size because of the irregular shapes of fossils (see Dunham 1962).

Type II limestones (designated as Microcrystalline Allochemical rocks) consist also of a considerable proportion of allochems, but here currents were not strong enough or persistent enough to winnow away the microcrystalline ooze, which remains as a matrix: sparry calcite is very subordinate or lacking simply because no pore space was available for it to form in. These rocks are equivalent texturally to the clayey sandstone or conglomerates, which also tend to have little cement. In these rocks the

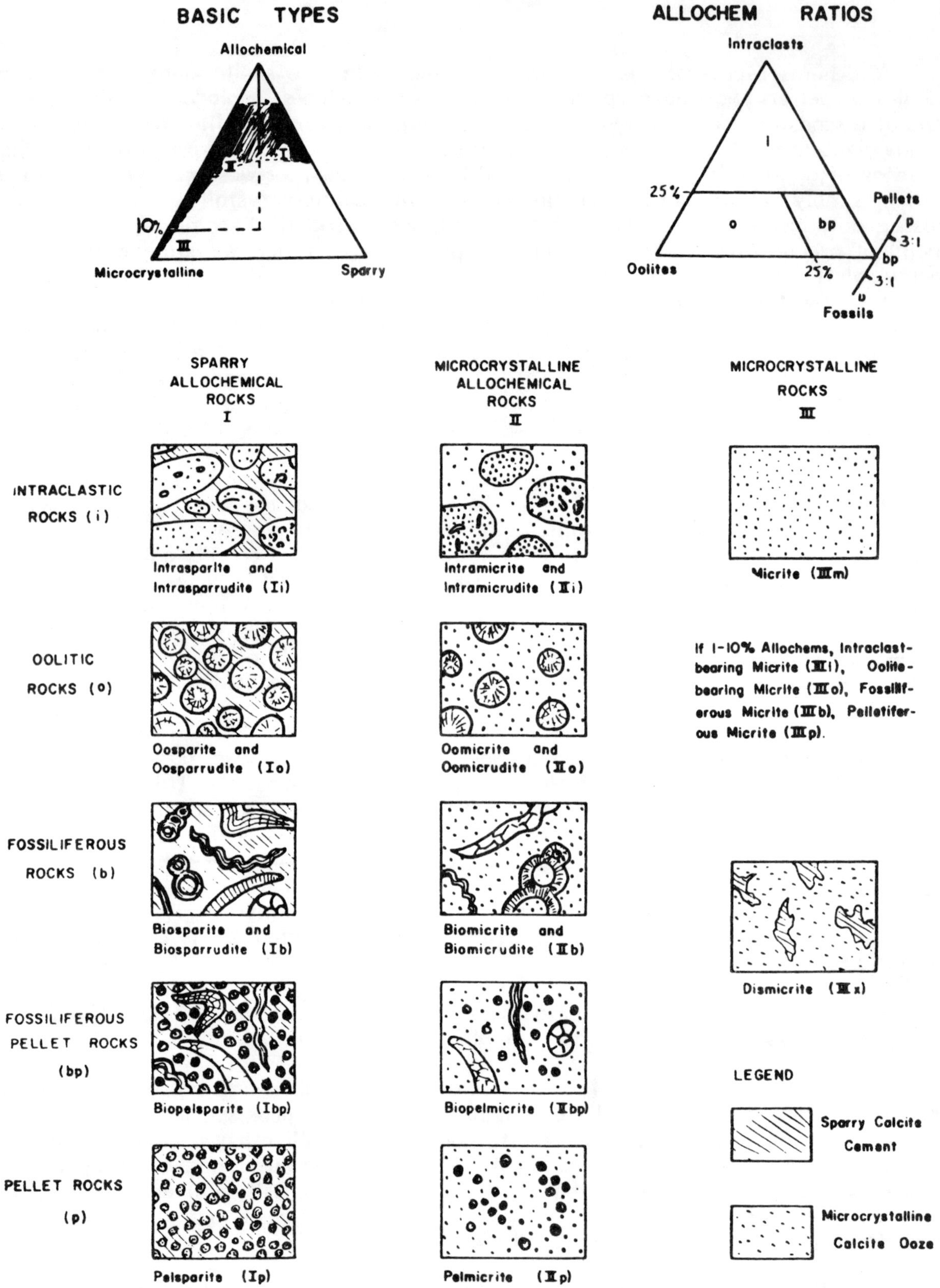

COMPOSITIONAL CLASSIFICATION OF LIMESTONES

restrictions of packing impose a certain maximum on the amount of allochems; yet there is no minimum, and Microcrystalline Allochemical rocks are found with percentages of allochems (intraclasts, oolites, fossils, or pellets) varying continuously from about 80 percent down to almost nothing. The reason for this is that microcrystalline ooze can form a rock in its own right (comparable to a claystone in the terrigenous series), and can accept any amount of allochem material that becomes mixed with it. Thus the boundary line between Microcrystalline Allochemical rocks and Microcrystalline rocks is entirely arbitrary, and has been set at 10 percent allochems.

Type I limestones indicate strong or persistent currents; type II limestones indicate weak, short-lived currents or a rapid rate of formation of microcrystalline ooze; and most limestones can be assigned readily to one or the other of these two classes because usually either sparry calcite or microcrystalline calcite is clearly dominant. In some rocks there are transitions, however, either because (1) washing is incomplete and the ooze is only partially removed, or (2) in some very fine grained pellet calcilutites, the pore spaces between pellets are so tiny that sparry calcite crystals are very minute, and can only with great difficulty be told from microcrystalline ooze. Transitional types can be designated by symbol I - II. Some of these types may also represent rocks in which the matrix has partially recrystallized.

Type III limestones (the Microcrystalline rocks) represent the opposite extreme from Type I, inasmuch as they consist almost entirely of microcrystalline ooze with little or no allochem material and no sparry calcite. This implies both a rapid rate of formation of microcrystalline ooze together with lack of strong currents. Texturally, they correspond with the claystones among the terrigenous rocks.

Some microcrystalline rocks have been disturbed either by boring organisms or by soft-sediment deformation, and the resulting openings are filled with irregular "eyes" of sparry calcite. Other beds of microcrystalline ooze have been partially torn up by bottom currents and rapidly redeposited but without the production of distinct intraclasts. These are considered as Disturbed Microcrystalline rocks, and a special symbol and rock term ("dismicrite") is used for them (see classification table).

Bioherm rocks (Cummings and Shrock, 1928), made up of organic structures growing _in situ_, are unique and place in a special class, "biolithite," Type IV. The dominant organism should be specified, e.g., coral biolithite, blue-green algal biolithite, etc. Of course, if these bioherms are broken up and redeposited the resulting rock is considered to be made up of intraclasts and falls in Type I or Type II depending on the interstitial material.

After the main division of limestones into Types I, II, or III it is most essential to distinguish whether the allochemical portion consists of intraclasts, oolites, fossils or pellets. In terrigenous sandstones, one wishes to know not only whether the rock has clay or not, but what the composition of the sand is; hence geologists recognize arkoses, phyllarenites and orthoquartzites, all of which types may or may not contain clay matrix. It is just as important to recognize the radically different allochem types in limestones, and the scheme for classification is presented in the table. The division lines between the groups are set at levels believed to reflect the significance of the constituent; for example, intraclasts are so important genetically, indicating as they do a tearing up of previously-deposited limestone and possibly indicating tectonic uplift, that a rock is called an intraclastic rock if it contains only 25 percent intraclasts, although it may have 60 to 70 percent fossils. Whether a rock is intraclastic, oolitic, biogenic, or pelletiferous is indicated by adding "i", "o", "b", or "p" to the symbol I or II or III, as in Sparry Intraclastic rocks (Ii), or Microcrystalline biogenic rocks (IIb).

So far, gross texture (whether Sparry Allochemical, Microcrystalline Allochemical, or Microcrystalline), and composition of allochems (whether intraclasts, oolites, fossils, or pellets) have been included in the classification, but nothing has been said about grain size of the allochems. If the allochems average coarser than 1 mm, the rock is a calcirudite (or dolorudite); if they lie between .0625 and 1 mm, the rock is a calcarenite or dolarenite; if finer than .0625 mm, calcilutite or dololutite. In determining the grain size name, only the size of the allochems is considered; percentage and crystal size of microcrystalline ooze or sparry calcite and grain size of terrigenous material is ignored. Thus a rock consisting of 20 percent brachiopod shells embedded in microcrystalline ooze and quartz sand is descriptively a calcirudite, just as much as a rock consisting of 80 percent limestone pebbles cemented by sparry calcite. Genetically, of course, there is a great difference between them and the writer seldom uses the term "biomicrudite".

In theory, the three fold size classification just given is valid; but in practice, rocks with allochems averaging in the calcilutite range are very rare. The only allochem rock types with representatives in this size class are pellet rocks or biogenic rocks, and in both of these the pellets or fossil fragments average no smaller than .04 or .05 mm, just barely under the limit of calcarenite; pellet rocks nearly always hover on the borderline between calcilutite and calcarenite (since the average pellet size is between .05 and .10 mm). Setting a new rock class apart on such an artificial and insignificant boundary seems to be an unnecessary complication, hence the writer has lumped these rare allochem calcilutites together with the calcarenites in the classification scheme. The only common calcilutites are the pure microcrystalline oozes, although many pellet rocks appear as calcilutites in the field and under a binocular microscope.

All the rock characteristics discussed above are combined in a single name, shown in the table and diagrammatically in the second figure. At first the writer used such cumbersome terms as "sparry intraclastic calcarenite" for intrasparite, "microcrystalline biogenic calcirudite" for biomicrudite, etc.; but these names, although self-explanatory, were too awkward to use in descriptions. As an alternative he thought of introducing locality terms, but the localities would be difficult to choose and the terms in themselves would be entirely meaningless and difficult to memorize. Finally, the writer decided to use single composite words, each portion of which referred to a specific rock characteristic: thus "intra" for intraclastic rocks, "oo" for oolitic ones, "bio" for biogenic types, and "pel" for pellet rocks. Whether the rocks were type I or type II was shown by the second part of the name, "spar" for those with sparry calcite cement and "micr" (pronounced with a short i, as in "mick") for those with microcrystalline ooze matrix. Type III limestones, almost entirely ooze, were designated "micrite" (pronounced "mick-rite") without any allochem prefix. Inasmuch as most limestones are calcarenites, no further syllable was added if the size fell in that category, and as exlained above, the rare and somewhat artificial calcilutites are lumped together with the calcarenites in this table. It is important to differentiate the calcirudites, though, and the word segment "rudite" may be added if the size falls in that class. Examples, together with the symbols used, are presented in the table.

Some rocks classified as Oosparite, Intramicrudite, etc. may have significant amounts of other allochems which do not appear in the name. These may be specified at the discretion of the worker, such as Fossiliferous Oosparite, Oolitic Intramicrudite, etc. Biogenic rocks, if composed largely of one or two types of organism, should always be described as Brachiopod biomicrudite, Gastropod biosparite, Foramcrinoid biomicrite, Oyster-echinoid biosparite, etc.

The writer has not yet adapted this classification to rocks in which recrystallization has been important; any such attempt would be quite premature until these rocks have been better studied and their importance evaluated. One recrystallized rock type appears to be of rather frequent occurrence in limestones, however. These rocks would be classified as micrites, type III, or biomicrites, type IIb, (i.e., nearly pure microcrystalline ooze, or fossils in an ooze matrix, respectively) were it not for the fact that the grains are still equidimensional and uniform in size but microcrystalline calcite is coarser than normal--average 5 to 15 microns instead of 2 to 5 microns. Because this relatively coarser material occupies large areas or makes up the entire specimen, it cannot have formed as a cement and probably represents aggrading recrystallization of a "normal" microcrystalline ooze matrix. These rocks the writer has designated as microsparite (corresponding to micrite), and biomicrosparite (corresponding to biomicrite), with symbols respectively RIIIm and RIIb.

Grain Size Scale for Carbonate Rocks

Carbonate rocks contain both physically-transported particles (oolites, intraclasts, fossils, and pellets) and chemically-precipitated minerals (either as pore-filling cement, primary ooze, or as products of recrystallization and replacement). Therefore the size scale must be a double one, so that one can distinguish which constituent is being considered (e.g. coarse calcirudites may be cemented with very finely crystalline dolomite, and fine calcarenites may be cemented with coarsely crystalline calcite). The size scale for transported constituents uses the terms of Grabau but retains the finer divisions of Wentworth except in the calcirudite range; it is most useful for limestones where transported particles are most obvious. For obviously allochemical dolomites, the terms "dolorudite", "dolarenite" and "dololutite" are substituted for these shown. The scale for authigenic constituents also follows the Wentworth divisions; it is most useful for dolomites, where transported particles are usually obliterated by replacement, and crystal size is one of the few describable characteristics. Most dolomites fall in the medium crystalline range. For symbolic abbreviation, use numbers as shown.

	Transported Constituents	Authigenic Constituents	
256 mm			256 mm
	Very coarse calcirudite	(9)	
64 mm			64 mm
	Coarse calcirudite	(8) Extremely coarsely crystalline	
16 mm			16 mm
	Medium calcirudite	(7)	
4 mm			4 mm
	Fine calcirudite	(6) Very coarsely crystalline	
1 mm			1 mm
	Coarse calcarenite	(5) Coarsely crystalline	
0.5 mm			
	Medium calcarenite		
0.25 mm			0.25 mm
	Fine calcarenite	(4) Medium crystalline	
0.125 mm			
	Very fine calcarenite		
0.062 mm			0.062mm
	Coarse calcilutite	(3) Finely crystalline	
0.031 mm			
	Medium calcilutite		
0.016 mm			0.016mm
	Fine calcilutite	(2) Very finely crystalline	
0.008 mm			
	Very fine calcilutite		
0.004 mm			0.004mm
		(1) Aphanocrystalline	
0.002 mm			
0.001 mm			0.001mm
		(0) Cryptocrystalline	

<u>Carbonate composition</u>. All of the rock types described above and listed in the table can occur either as limestone or dolomitized limestone, and some may occur as primary dolomite. If the rock is a limestone, the rock name (e.g. oosparite or pelmicrite) is used unmodified, intrasparrudite, for example. If the rock contains over 10 percent replacement dolomite "dolomitized" is prefixed to the main rock name (e.g. Dolomitized Oosparite, or Dolomitized Pelmicrite). If the dolomite is of uncertain origin, the term "dolomitic" is suggested. In case the rock is a "primary" dolomite, use Dolomite Intramicrudite, etc.

Limestones that have been completely replaced by dolomite offer considerable difficulty since in many cases the original structure is partly obliterated. Fine-grained clastic particles such as pellets or finely-broken fossils are especially prone to vanish. Likewise, one does not know the original proportion of micro-crystalline ooze versus sparry calcite cement. In such cases it is very difficult if not impossible to allot a dolomite to either classes I, II, or III; if ghost oolites, fossils, intraclasts, or pellets are present, that fact can be indicated by a modifier. The crystal size of these rocks is a very important characteristic and should be shown by the following terms and symbols:

Cryptocrystalline	0	under		.001	mm
Aphanocrystalline	1	.001	-	.0039	mm
Very finely crystalline	2	.0039	-	.0156	mm
Finely crystalline	3	.0156	-	.0625	mm
Medium crystalline	4	.0625	-	.25	mm
Coarsely crystalline	5	.25	-	1.00	mm
Very coarsely crystalline	6	1.00	-	4.00	mm
Extremely coarsely crystalline	7	over		4.00	mm

Examples of replacement dolomite rock names are Medium Crystalline Intraclastic Dolomite, Finely Crystalline Biogenic Dolomite, or for a rock with no visible allochems, Coarsely Crystalline Dolomite.

Terrigenous Admixture. So far we have ignored the content of terrigenous particles. If the rock contains over 50 percent terrigenous material, it is a Terrigenous rock and not further considered here. If it contains less than 10 percent terrigenous material, it is a Pure Chemical rock and the terrigenous content is so low that it is not mentioned in the classification.

However, if the rock contains between 10 and 50 percent terrigenous material, that is regarded as important enough to be mentioned in the name and in the classification symbol. These rocks as a class are known as Impure Chemical rocks; a specimen of this type is classified just as previously described (i.e., as a biomicrite, oosparite, etc.), but to identify it as an Impure Chemical rock the terrigenous admixture is described.

The following list shows examples of this usage:

Clayey Biopelmicrite
Silty Coarsely Crystalline Dolomite
Sandy Dolomicrite
Sandy Dolomitized Intrasparite

The Classification used here necessarily is determined by relative rates of formation of each constituent, not on absolute rates. Thus an abundance of terrigenous material in a limestone may mean (1) that uplift or proximity of the source area caused a more rapid influx of detritus; (2) a change of conditions in the depositional basin suppressed chemical activity, so that terrigenous minerals accumulated by default; or (3) current velocities were such as to concentrate terrigenous material of a certain size in preference to allochemical material of different size.

Abundance of the types. Some remarks may be made as to the relative abundance of these various rock types in the stratigraphic section as a whole. These observations are based on examination of several thousand thin sections of carbonate rocks from many regions.

Intraclastic rocks usually have a sparry calcite cement, inasmuch as currents that are strong enough to transport fairly large carbonate rock fragments are also usually capable of washing away any microcrystalline ooze matrix. Thus, rocks of type Ii (intrasparite) are common, whereas type IIi (Intramicrite) is relatively rare. Texturally, intraclastac rocks are about equally divided between calcirudites and calcarenites.

Oolite rocks with their high degree of sorting imply fairly vigorous current action, therefore Oosparite (type Io) is much more abundant than Oomicrite (type IIo). Texturally these rocks are nearly always calcarenites, although in some specimens the oolites may reach an average diameter larger than 1 mm, in which case the rock would be a calcirudite. Pisolite rocks might be classified as Pisosparite.

Biogenic rocks may occur just as frequently with a microcrystalline ooze matrix (Biomicrite, type IIb) as with a sparry calcite cement (Biosparite, type Ib). Chalk is usually foraminiferal biomicrite or foraminiferal micrite.

Type IIb rocks indicate either that the fossils are sedentary or else that currents were calm in the depositional area and the microcrystalline ooze did not get winnowed out from the shell material. Type Ib rocks indicate deposition under vigorous current action where the microcrystalline material was washed away. Note that both intraclastic rocks and oolitic rocks require vigorous current action in order to form, thus are usually -sparites; while biogenic rocks do not, hence may have either microcrystalline matrix or sparry calcite cement. Biogenic rocks are most commonly calcirudites or calcarenites, although calcilutites occur sometimes if the fossils are very fine-grained foraminifera.

Pellet rocks are quite common, but are often mistaken in the field or even under the binocular microscope for microcrystalline rocks. Usually they have a sparry calcite cement, thus belong in type Ip (Pelsparite) although sometimes they have a microcrystalline matrix (type IIp, Pelmicrite). Texturally they are borderline between very fine calcarenites or coarse calcilutites, but they are all of such uniform size that the writer designates them all as Pelsparite or Pelmicrite regardless of the precise average diameter of the allochems.

Microcrystalline rocks, type III (Micrites, type III, or Dismicrites, type IIIX) occur frequently in the section. They quite often contain more than 10 percent clay, thus type TcIII (Clayey Micrite) is common; Fossiliferous Micrite (IIIb) is another rather frequent type.

Biolithites (type IV) are uncommon but interesting. A rock should not be called this unless it consists of organisms essentially in situ and forming a rigid framework; it should not be applied to broken and redeposited fragments. Coral biolithite, Blue-Green-Algal biolithite, Red-Algal biolithite, and bryozoan biolithite are most common. These may form either mounds or more blanket-like bodies.

Genetic Significance. A host of papers and books are now available on this subject (e.g. Bathurst, 1971; Milliman, 1974; Wilson, 1975).

In general, type I limestones (Sparry Allochemical Limestones, or -sparites for short) are those limestones deposited in environments of vigorous winnowing acton and pretty efficient sorting; they often show good bedding or cross-bedding, close packing and good orientation of allochems. Thus they form as eolianites beaches, shallow neritic sediments, submarine raised banks, swift tidal channels, barrier bars, etc. To these sediments the concept of textural maturity can be fruitfully applied (this idea stimulated by R. J. Dunham and C. B. Thames). Of course all type II limestones (-micrites) would be classed as immature, and the -sparites would be submature or better. The poorly sorted -sparites would be submature, well sorted ones mature, and sorted and rounded ones supermature. The maturity concept would be without much significance for pelsparite and oosparite, which are virtually by definition well sorted and well rounded. Intrasparites could be divided into well-sorted versus poorly sorted types, but intraclasts are almost always rounded so that "supermature" would have no meaning. The most cogent application would be to the biosparites. Supermature biosparites (finely broken and heavily abraded and rounded fossil fragments get ground to bits by wave action to form rounded, polished coquinas. Such supermature pelecypod biosparites formed a semicircular beach coquina around the north flank of Pilot Knob, a Cretaceous submarine volcano near Austin (R. White). Often, the well-sorted (mature) ones are made up of one fossil type, e.g., pelecypod fragments, crinoid columnals, forams, or fusulinids; the poorly sorted (submature) ones are made up of diverse mixtures of fossils of inherently different size, e.g. forams plus crinoids, or brachiopods plus bryozoans plus fusulinids. Thus sorting is believed, as in terrigenous rocks, to be largely a function of source (i.e. available types of fossils), with environment (i.e. strength of waves and currents, or water depths) playing a modifying, but not controlling role. Sorting is best in shallower waters or on the beach itself, and gets worse offshore. Thus in moving from a beach surf zone into deeper waters one might theoretically expect the sequence to go from supermature biosparite (well rounded and sorted) to mature (sorted but only slightly rounded) to submature (unsorted, fragments not rounded but possibly broken) to biomicrite (immature, with unsorted, broken or whole fossils) and finally micrite. This sequence would undoubtedly be modified by types of fossils available and wave energy of the coast. See 1962 AAPG Mem #1 and 1964 JG (Folk & Robles) for details.

In some areas, shoal regions and littoral zones are characterized by algal, coral, or other types of reefs. Most reef-forming organisms live in the shallow waters where there is still sunlight available for photosynthesis of the algae, which make up the bulk of living tissue on most reefs (Odum). Reef rocks (biolithites) are exceedingly complex, because of (1) abundant life and many different types of fossils, making rock petrographically varied, especially the algae (there is probably as much complexity and variety of external form and internal structure in algae as there is in all other fossils put together), (2) the fantastic growth forms assumed by many of the reef organisms, with weird branching, bushy or encrusting habits, (3) the reef rock is riddled with cavities which may be empty (and later filled with spar in fibrous or mosaic forms), or cavities may be partly filled with terrigenous sand, intraclasts, lime mud, pellets, organisms that lived in the cavity, and inswept broken fossil debris. Add to this the fact that reefs are prone to patchy recrystallization (because of the high aragonite content) and one can see that a thin section will be of the utmost complexity. Reefs, although they usually thrive in high-energy zones, act as baffles to the current, and the water movement down in the crevices and pockets between branching organisms may be relatively calm even in a surf zone, just as the interior of a forest may be calm in a windy day. Hence reefs are associated with much micrite and fine debris which filters down between the branching growths and cannot be dislodged. In fact, stromatolitic reefs, probably formed by blue-green algae, are largely micrite, presumably trapped by the slimy algal mats and then firmly bound together by the filaments.

Rocks of type II and type III (either pure micrite or rocks with a micrite matrix) indicate formation in areas of ineffective winnowing and calm currents. These can form in four important ways: (1) in protected lagoons, in which the water is very shallow, perhaps not more than a few feet deep; (2) in broad, shallow platforms on the lee side of barriers (e.g. west side of Andros Island, Bahamas), where the great width of the platform prevents any permanent removal of lime mud and it is merely shifted around; (3) in moderately deep waters in geosynclines (probably down to a few hundreds of feet); (4) as open-ocean Chalks; (5) as Ginsberg has shown, lime mud may accumulate locally around organic baffles (marine grasses, coral or algal growths) even in fairly high-energy environments. To tell these micrite-rich environments apart is often a difficult problem as they may represent very shallow or moderately deep areas. Several criteria are useful. Among the best criteria is the fossil content. For example, clams or oysters in micrite would indicate a shallow, perhaps lagoonal area, whereas an exclusively pelagic fauna of small forams would probably be indicative of deeper water

CARBONATE TEXTURAL SPECTRUM

	OVER 2/3 LIME MUD MATRIX				SUBEQUAL SPAR & LIME MUD	OVER 2/3 SPAR CEMENT		
Percent Allochems	0-1 %	1-10 %	10-50%	OVER 50%		SORTING POOR	SORTING GOOD	ROUNDED & ABRADED
Representative Rock Terms	MICRITE & DISMICRITE 1	FOSSILI-FEROUS MICRITE 2	SPARSE BIOMICRITE 3	PACKED BIOMICRITE 4	POORLY-WASHED BIOSPARITE 5	UNSORTED BIOSPARITE 6	SORTED BIOSPARITE 7	ROUNDED BIOSPARITE 8
1959 Terminology	Micrite & Dismicrite	Fossiliferous Micrite	Biomicrite			Biosparite		
Terrigenous Analogues	Claystone		Sandy Claystone	Clayey or Immature Sandstone		Submature Sandstone	Mature Sandstone	Supermature Sandstone

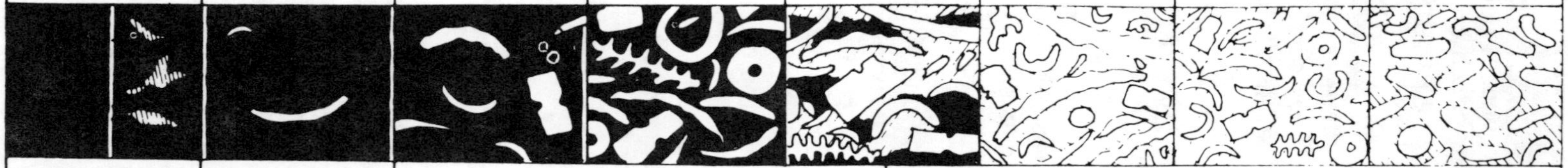

■ LIME MUD MATRIX

□ SPARRY CALCITE CEMENT

VOLUMETRIC ALLOCHEM COMPOSITION			LIMESTONES, PARTIALLY DOLOMITIZED LIMESTONES, AND PRIMARY DOLOMITES (see Notes 1 to 6)					REPLACEMENT DOLOMITES[7] (V)	
			>10% Allochems ALLOCHEMICAL ROCKS (I and II)		<10% Allochems MICROCRYSTALLINE ROCKS (III)		UNDISTURBED BIOHERM ROCKS (IV)	Allochem ghosts	No Allochem Ghosts
			Sparry calcite cement > Microcrystalline ooze Matrix	Microcrystalline Ooze Matrix > Sparry Calcite Cement	1-10% Allochems	<1% Allochems			
			SPARRY ALLOCHEMICAL ROCKS (I)	MICROCRYSTALLINE ALLOCHEMICAL ROCKS (II)	Most Abundant Allochem			Evident Allochem	
>25% Intraclasts (i)			Intrasparrudite (Ii:Lr) Intrasparite (Ii:La)	Intramicrudite* (IIi:Lr) Intramicrite* (IIi:La)	Intraclasts: Intraclast-bearing Micrite* (IIIi:Lr or La)	Micrite (IIIm:L); if disturbed, Dismicrite (IIImX:L); if primary dolomite, Dolomicrite (IIIm:D)	Biolithite (IV:L)	Finely crystalline Intraclastic Dolomite (Vi:D3) etc.	Medium Crystalline Dolomite (V:D4)
<25% Intraclasts	>25% Oolites (o)		Oosparrudite (Io:Lr) Oosparite (Io:La)	Oomicrudite* (IIo:Lr) Oomicrite* (IIo:La)	Oolites: oolite-bearing Micrite* (IIIo:Lr or La)			Coarsely Crystalline Oolitic Dolomite (Vo:D5), etc.	Finely Crystalline Dolomite (V:D3)
	<25% Oolites; Volume Ratio of Fossils to Pellets	>3:1 (b)	Biosparrudite (Ib:Lr) Biosparite (Ib:La)	Biomicrudite (IIb:Lr) Biomicrite (IIb:La)	Fossils: Fossiliferous Micrite (IIIb: Lr, La, or Ll)			Aphanocrystalline Biogenic Dolomite (Vb:D1). etc.	
		3:1 - 1:3 (bp)	Biopelsparite (IIbp:La)	Biopelmicrite (IIbp:La)	Pellets: Pelletiferous Micrite (IIIp:La)			Very Finely Crystalline Pellet Dolomite (Vp:D2), etc.	etc.
		<1:3 (p)	Pelsparite (Ip:La)	Pelmicrite (IIp:La)					

far from shore. Another very good criterion is the lithology of associated beds. If the micrite occurs interbedded with high-energy calcarenites and calcirudites, cemented with sparry calcite and showing evidences of current action, then the micrite is very likely of shallow water origin; the alternation might represent a fluctuation between lagoon and barrier-bar or beach environments. If, on the contrary, monotonous thicknesses of micrite occur with no interbedded -sparites, a deeper water origin is implied. Burrowed or churned up dismicrites seem to be characteristic of lagoonal or shallow-water sediments, as are micrites that have been torn up and reworked to form intraclasts. If the climate was semiarid with development of evaporitic tendencies, the lagoonal micrites might be associated with beds of more saline indicators such as replacement dolomite, or, if restricted circulation, primary dolomite, anhydrite, gypsum, celestite, etc. If the climate was humid, influxes of clay may be associated with lagoonal micrites because of river drainage into bays and lagoons.

In many beds of biomicrite and fossiliferous micrite, the fossils are badly broken; how does this happen in supposedly calm-water environments? Sometimes these beds would represent a normal sequence from shoreline well-rounded biosparites into more poorly sorted biosparites made of broken but non-rounded shells, to biomicrites and finally micrites in deeper or calmer waters. But often the shells are broken not by wave or current action but by the activities of predator organisms, which crunch up the shells in search of food and in some cases pass the debris through their digestive tracts. Many beds of micrite show badly broken and randomly oriented shells distributed patchily throughout the rock, often arranged along vague "flow-lines." These probably represent a lime mud riddled with burrowers and scavengers, which break the shells and churn them up with micrite.

Beds of intramicrite and oomicrite might be formed by a similar mechanism. A bed of lime mud might be laid down, followed by deposition of a well-sorted layer of intraclasts or oolites. Before the rocks have a chance to get cemented, the whole sediment might be churned up by burrowers so that the allochems would get mixed with micrite. In these rocks the allochems often show a patchy, swirled distribution.

The significance of color in limestones is not too well known. Ordinarily, in well-oxygenated (shallow and wave-agitated) waters, bacteria thrive and eat up decayed organic matter, resulting in light-colored rocks. In waters of poor oxygenation, either restricted, stagnating lagoon or deeper offshore marine waters, reducing conditions dominate and organic matter accumulates faster than bacteria can remove it, and dark, pyritic limestones result. The Buda limestone in central Texas grades from white micrite in the nearshore facies to dark gray micrite in deeper wates farther offshore (Hixon). Many deep geosynclinal limestones are dark colored, but so are many shallow lagoonal limestones.

Notes to Classification Table:

* designates rare rock types.

1. Names and symbols in the body of the table refer to limestones. If the rock contains over 10 percent replacement dolomite, prefix the term "dolomitized" to the rock name, and use DLr or DLa for the symbol (e.g. Dolomitized Intrasparite, Ii:DLa). If the rock contains over 10 percent dolomite of uncertain origin, prefix the term "dolomitic" to the rockname and use dLr or dLa for the symbol (e.g. Dolomitic biomicrite IIb:dLa). If the rock is a primary dolomite, prefix the term "primary dolomite" to the rock name, and use Dr or Da for the symbol (e.g.

Primary Dolomite Intramicrudite, IIi:Dr). Instead of "Primary Dolomite Micrite" the term "Dolomicrite" may be used.

2. Upper name in each box refers to calcirudites (median allochem size larger than 1.0 mm), and lower name refers to all rocks with median allochem size smaller than 1.0 mm. Grain size and quantity of ooze matrix, cements, or terrigenous grains are ignored.

3. If the rock contains over 10 percent terrigenous material, prefix "Sandy", "Silty", or "Clayey" to the rock name, and "Ts", "Tz", or "Tc" to the symbol depending on which is dominant (e.g. Sandy Biosparite, TsIb:La, or Silty Dolomitized Pelmicrite, TzIIp:DLa). Glauconite, collophane, chert, pyrite, or other modifiers may also be prefixed.

4. If the rock contains other allochems in significant quantities that are not mentioned in the main rock name, these should be prefixed as qualifiers immediately before the main rock name (e.g. Fossiliferous Intrasparite, Oolitic Pelmicrite, Pelletiferous Oosparite, or Intraclastic Biomicrudite). This can be shown symbolically as Ii(b), Io(p), IIb(i), respectively.

5. If one or two types of fossils are dominant, this fact should be shown in the rock name (e.g. Pelecypod Biosparrudite, Crinoid Biomicrite, Brachiopod-Bryozoan Biosparite, etc.).

6. If the rock was originally microcrystalline and can be shown to have recrystallized to microspar (5 to 10 microns, clear calcite) the terms "microsparite", "biomicrosparite" etc. can be used instead of "micrite" or "biomicrite".

7. Specify crystal size as shown in the examples.

DIAGENESIS

Diagenesis, the conversion of carbonate sediment into rock, takes place by a multitude of processes, described by a recent avalanche of papers (see Folk 1974 JSP). Three processes go on: (1) solution of the more unstable minerals, particularly aragonite; (2) filling of pores by precipitated minerals of all kinds, but most abundantly calcite; (3) alteration of original minerals to newer ones, stable under changed conditions, by a complex of processes such as expulsion of Mg from mg-calcite, inversion of aragonite to calcite, recrystallization of fine calcite to coarser calcite, or replacement by dolomite (or silica, pyrite, etc.).

Most diagenesis takes place near the contact zones between two or three of the following phases: air, fresh water, sea water, and sediment. Diagenesis and precipitation of cements is particularly active where two solutions of different composition, temperature, CO_2 content, etc. mix. If Karl Marx will forgive us, we may state the Geochemist Manifesto: "Waters of the World, Unite! You have nothing to lose but your ions." The accompanying diagram shows common environments of carbonate diagenesis.

The minerals that form upon diagenesis are obviously controlled by chemistry of the environment. The main features can be explained by making a plot of the salinity vs. Mg/Ca ratio of the diagenetic solutions. Two ideas are fundamental:

(1) Calcium Carbonates. The presence of Mg/Ca ratios over about 2:1 forces $CaCo_3$ to be precipitated as either aragonite or magnesian-calcite. Mg++ poisons sideward growth of these crystals, so they tend to be fibers or steep-faced rhombs. Only when Mg/Ca ratios drop below about 2:1 can equant, sparry calcite form. This means that in sea water, hypersaline brines, beach-rock, etc. with Mg/Ca ratios varying from 3:1 to as much as 100:1 in some sabkhas, that aragonite and magnesian calcite fibers or micrite form. In fresh surface waters, and in most subsurface brines, where Mg/Ca ratios generally are 1:10 to 1:2, equant to bladed sparry calcite forms.

(2) Dolomite is difficult to crystallize because of the precise ordering required to form the alternating sheets of Ca and Mg. Thus it forms best when crystallization is slow, precipitating from dilute solutions. Dolomite formation is also favored by a high Mg/Ca ratio; in hypersaline solutions, dolomite forms at Mg/Ca over 5 or 10:1, but in "fresh" waters it can form at Mg/Ca values as low as 1:1. Thus the dolomite boundary forms a sloping line on this graph: Its formation is favored by reducing salinity, or by increasing the Mg/Ca ratio. Fresh water dolomite, because of its slow crystallization, is characteristically euhedral and "limpid" (water-clear with sharp, mirror-like facets). Limpid dolomite is much more slowly soluble than ordinary dolomite. Hypersaline dolomite is in very tiny crystals, poorly developed.

The following are the main diagenetic realms:

A. Saline waters

1. Beachrock. Cementation mostly by evaporating sea water in inter-tidal zone, at the sea/sediment/air contact zone. Salinity normal marine to somewhat hypersaline, Mg/Ca ratio about 3:1 or a little higher. Aragonite and Mg-calcite cements, fibrous or micritic.

2. Sabkha, or evaporitic salt flat. Salinity 5-10 times that of normal sea water, Mg/Ca from 5:1 to 100:1. Fibrous to micritic aragonite or Mg-calcite; aphanocrystalline dolomite if Mg/Ca exceeds 5 or 10:1; also evaporite minerals.

3. Subtidal cement near the sea water/sediment interface. Salinity and Mg/Ca ratio same as normal sea water. Physically stable substrate the main key, to allow slow crystallization of cements. Includes reef cement, shallow submerged shelf (Persian Gulf), and Ocean-floor cements. Mainly micritic and fibrous Mg-calcite and aragonite; some dolomite and rare calcite (by very slow crystallization).

B. Meteoric waters

1. Surface fresh waters (lake, creeks). Salinity about 1% that of sea water; Mg/Ca ordinarily 1:10 to 1:2. Calcite the chief form; in waters of low ion strength (stream) calcite crystallizes as rhombs. In unusual lakes with high Mg/Ca ratio, Mg-calcite, aragonite and even dolomite form, even in waters of low salinity.

2. Shallow soils (air/sediment contact). Calcite in caliche can form micrite or rhombs. Dolomite reported where high-Mg waters present.

3. Vadose zone in calcarenites. Fresh water, very low Mg/Ca ratio. Sparry calcite, often minute rhombohedra.

4. Phreatic fresh-water zone. Salinity very low, Mg/Ca also generally low (1:10 to 1:3). Sparry calcite, often in large, poikilotopic crystals, extending to edge of pore without intervening crust. Most precipitation probably near water table (subsurface air/water contact). Upon weathering the tendency is to flush sediments with waters with a very low Mg/Ca ratio. Removal of Mg from the rocks allows recrystallization of micrite to microspar or pseudospar, and replacement of dolomite by sparry calcite (dedolomitization).

C. Subsurface waters

1. Deep subsurface waters, largely of meteoric derivation. Salinity rather low, Mg/Ca typically 1:6 to 1:2. Sparry calcite, generally anhedral. Some subsurface waters have Mg/Ca near 1:1, and here dolomite can form, limpid if crystallized from dilute solutions.

2. Deep subsurface waters, mainly of marine derivation. Salinity ranges up to several times sea water; Mg/Ca usually low, however, 1:4 to 1:2. Mg is selectively removed from subsurface connate brines because it is snatched out by clay minerals and by formation of dolomite. Removal of Mg from sea water means that Mg-poisoning no longer hampers calcite growth, and sparry calcite mosaic can form.

3. It is suspected that the greatest precipitation of subsurface carbonate should take place in the meteoric salt/water contact zone, e.g., near the base of the fresh water lens.

Dolomite formed rapidly at high salinities in sabkha environments tends to be in very tiny crystals and may be poorly ordered. Dolomite formed by fairly concentrated solutions in the shallow subsurface (so-called reflux) is coarser, but rather dirty with lots of inclusions. Dolomite formed by very slow crystallization from dilute solutions, at Mg/Ca ratios near the stoichiometric 1:1, tends to be "limpid", with perfect crystal faces and water-clear (Folk & Siedlecka, 1974). The easiest way to obtain dolomite is to reduce the salinity of a marine or hypersaline water by mixing in some meteoric water--such as in a schizohaline environment that fluctuates rapidly between super-high and low salinities, e.g., a hypersaline lagoon flooded by hurricane rains. Dilution drops salinity radically but maintains a high Mg/Ca ratio, dropping the water composition solidly into the dolomite field.

In surface waters of very low Mg/Ca ratio, Mg ions are removed from the rocks. This results in replacement of dolomite by calcite (dedolomitization), and in recrystallization of micrite to microspar or even pseudospar.

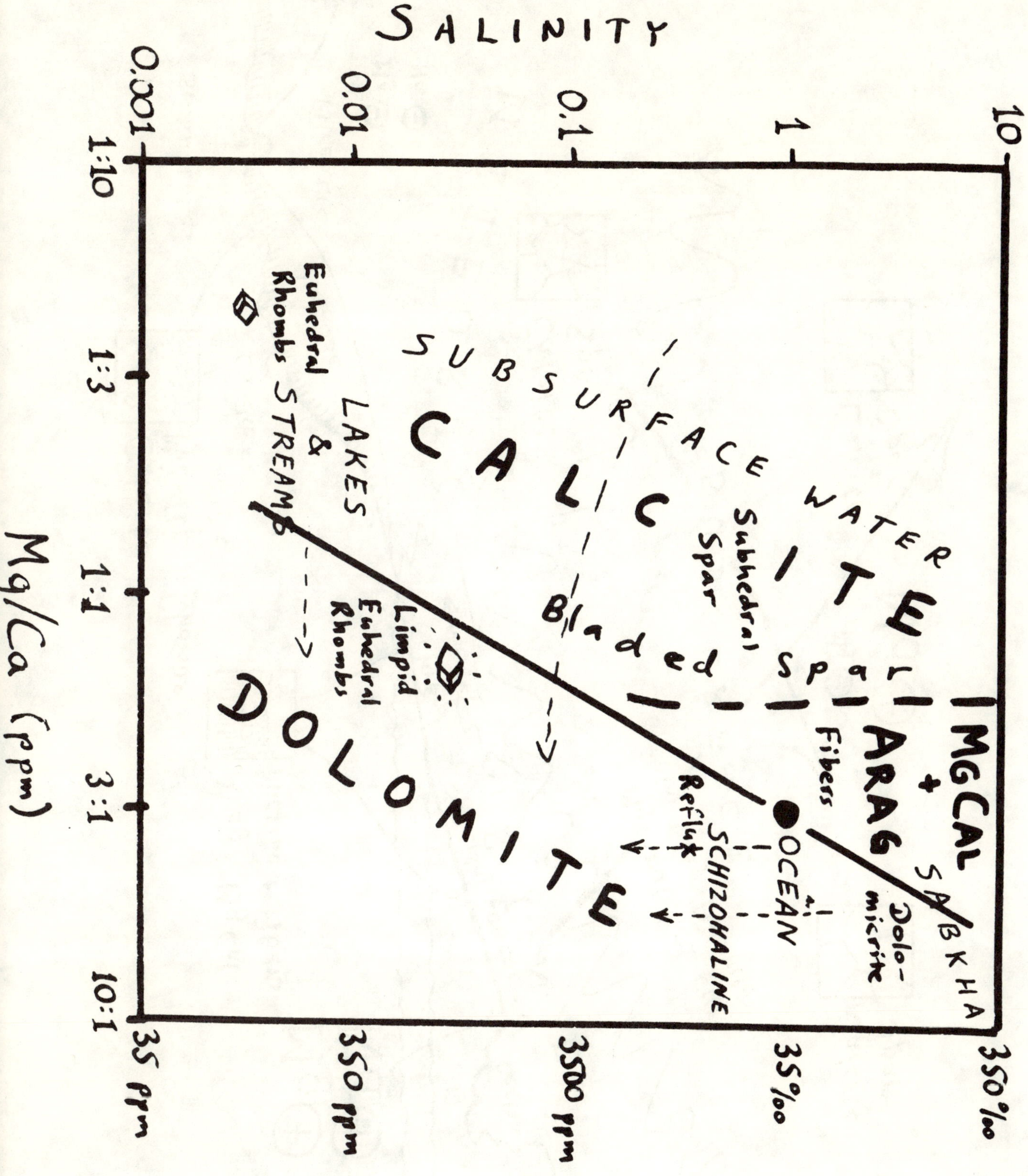
SALINITY
0.001
0.01
0.1
1
10
1:10
1:3
1:1
3:1
10:1
Mg/Ca (ppm)
35 ppm
350 ppm
3500 ppm
35‰
350‰
SUBSURFACE WATER
CALCITE
Subhedral Spar
Bladed Spar
LAKES & STREAMS
Euhedral Rhombs
Limpid Euhedral Rhombs
DOLOMITE
MG CAL + ARAG
Fibers
SABKHA
Dolo-micrite
OCEAN
SCHIZOHALINE
Reflux

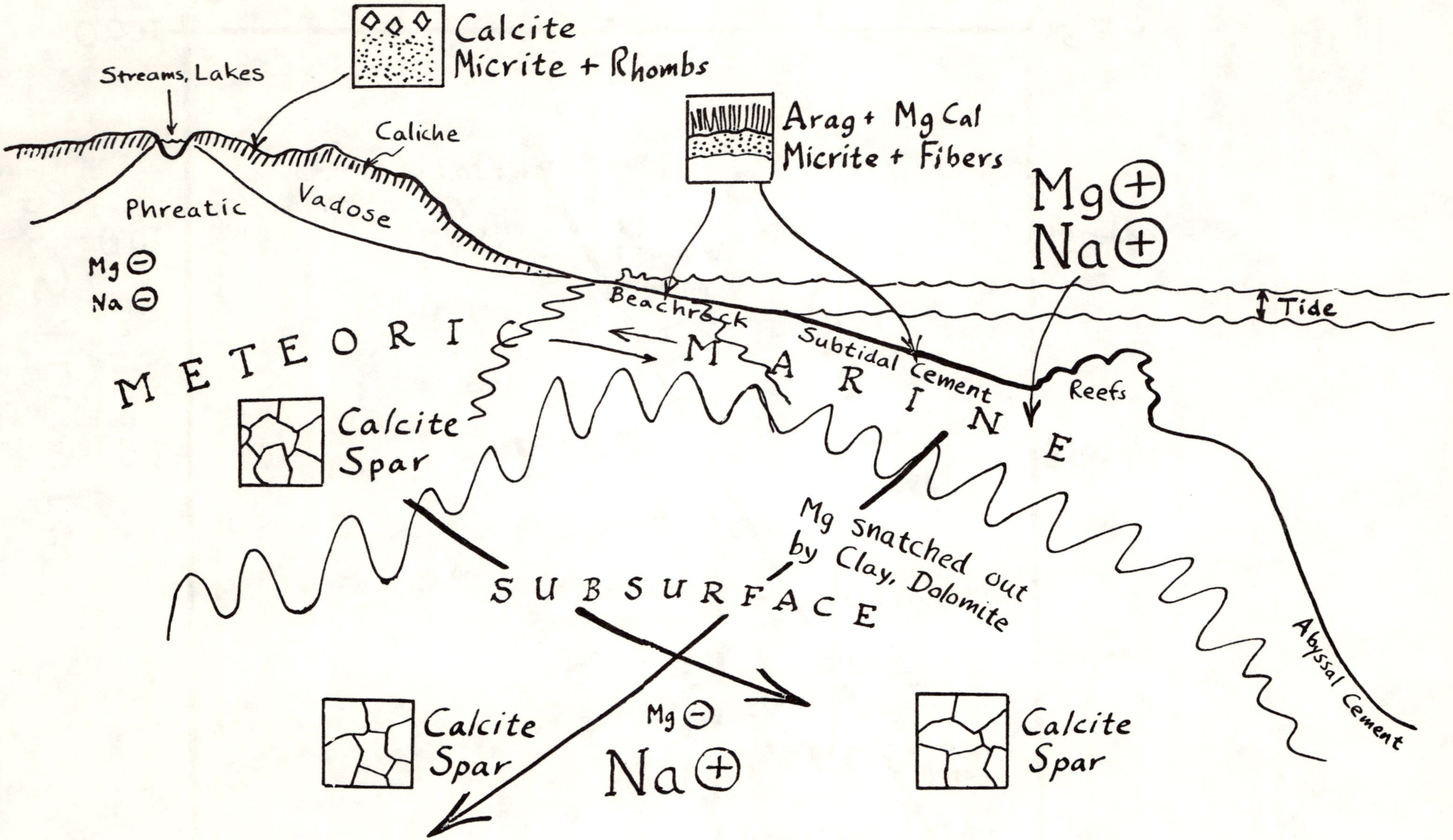
Calcite
Micrite + Rhombs
Streams, Lakes
Caliche
Arag + Mg Cal
Micrite + Fibers
Phreatic
Vadose
Mg ⊖
Na ⊖
Mg ⊕
Na ⊕
Beachrock
Tide
Subtidal Cement
Reefs
M E T E O R I C
M A R I N E
Calcite
Spar
Mg Snatched out
by Clay, Dolomite
S U B S U R F A C E
Abyssal Cement
Calcite
Spar
Mg ⊖
Na ⊕
Calcite
Spar

Recrystallization, Inversion and Neomorphism

Carbonates, more than most other rocks, are susceptible to alteration because of the highly reactive and unstable nature of many of their original constituents, particularly aragonite and High-magnesian calcite. Reaction may take the form of (1) total removal of the constituent, leaving an open cavity which exists for considerable time, and later filling by another mineral. This process, recognized as early as 1879 by H. C. Sorby, is a common one for aragonitic fossils whose molds are later filled by sparry calcite. Best evidence for this is partial collapse of the rim or "micrite-envelope" of the original shell (Bathurst, 1963). (2) Replacement of the original constituents by another mineral of grossly different composition, with no significant cavity developed; the host and guest minerals always remain in close contact, of course with a thin liquid film between to conduct the ions in and out of the system. This is the metasomatism of Lindgren, and is illustrated by dolomitization of a calcitic or aragonitic limestone, or replacement by chert, pyrite, etc. Replacement is usually volume for volume, indicating some addition or subtraction of ions from the system, (3) Inversion from one polymorph of a mineral to another, gross chemical composition remaining essentially constant--e.g. change from orthorhombic aragonite to rhombohedral calcite. This can happen either by (a) a migrating liquid-film causing simultaneous solution of the old and precipitation of the new as in ordinary replacement, or (b) a switching of the positions of the ions in the crystal lattices without the presence of liquids or long-distance export-import of ions. The liquid film process is probably much more important in ordinary carbonate rocks. Evidence is the preservation or original structure, e.g. organic laminae of the original aragonitic shell being preserved in the calcite pseudomorph (Hudson; Nelson; Sorby). (4) True recrystallization, where the original and final phases are the same mineralogically but merely differ in crystal form, size, or orientation. An example would be fibrous calcite converting to microcrystalline calcite, or microcrystalline calcite going to sparry calcite. Again, thin liquid films are probably present between the older and newer generations of crystals. This may be caused by strain (like the ordinary recrystallization--or as termed here "strain" recrystallization, of the metallurgist); or may be driven by other forces such as surface energy.

Other processes are sometimes encountered. For example, calcite in replacing an aragonitic shell may have a thin "no-man's land" of powdery carbonate, or even microporosity between the original and the new minerals (Schlanger 1963), a situation intermediate between solution-cavity-fill and inversion. Or High-magnesian calcite may expel Mg ions to leave relatively pure calcite of form and texture microscopically identical with the original.

Processes (3) and (4) above--inversion and true recrystallization may be lumped under the term "neomorphism" (Folk 1965) which simply means that a mineral has the same gross composition (ignoring changes in trace elements, isotopes, etc.) but merely has a new form--crystal size, shape or orientation--differing from the original. It is useful when referring collectively to inversion and recrystallization, or when the exact process is not known.

Neomorphism (and its more specific processes, inversion and recrystallization) usually result in increase of crystal size (aggrading recrystallization), such as when a carbonate mud goes to sparry calcite. Occasionally, the crystal size decreases (degrading recrystallization), as when oolites or fossils degrade to micrite, but this porcess is much more rare.

Aggrading neomorphism may be either porphyroid or coalescive. In porphyroid neomorphism (or porphyroid recrystallization, porphyroid inversion) a few large crystals grow at the expense of a static matrix, thus a porphyroid fabric with large and small crystals is in existence during intermediate stages of the process. In coalescive neomorphism, most of the crystals are either growing or being consumed, i.e. the whole mass is changing like the growth of bubbles in a soap foam; a uniform crystal size is maintained throughout the mass, though the average crystal size gradually grows larger. Porphyroid neomorphism appears to be a much more common process, but coalescive neomorphism may be the mechanism by which microspar is formed; it is usually very equigrained. The importance of calcite-to-calcite recrystallization is an unsolved point. In Folk's opinion this occurs importantly in some localities and at some stratigraphic horizons, but is overall volumetric importance is minor. Recrystallization of micrite to 5-15 micron microspar is rather common, however.

Recrystallized allochems can be recognized if they consist almost entirely of mosaic spar; some fossils and intraclasts are thus affected. Recrystallization of micrite matrix should yield allochems floating far apart in a "sea" of spar, like plums in a pudding, far too loosely packed to fit the requirements of packing; it should also produce isolated patches of recrystallization spar in homogeneous micrite beds. Such rocks are quite rare. Usually, where allochems are closely packed they have a sparry calcite cement (like close-packed, well-sorted sandstones); where allochems are loosely packed, they lie in a micrite matrix (like sand grains in a shale). The overwhelming dominance of these two types argues that most spar is a pore-filling cement, not the result of recrystallization. The abundance of well-sorted, now-porous calcarenites in recent sediments indicates that well-sorted, calcite-cemented calcarenites should be abundant in the geologic column.

Recrystallization can be proven if spar can be shown to transgress allochems and matrix more or less indiscriminately. The criteria for proving recrystallization are the same as those for replacement. Bathurst (1958) has porposed some fabric criteria for recognition of recrystallization, and Folk (1965) has a few comments. Longman and Mench (1978 Sed. Geol.) have a brilliant SEM study on fresh-water neomorphism of carbonates.

Dolomites

Dolomites are classified as shown in the preceding pages. Most dolomite is of replacement origin, and replacement usually seems to have occured while the limy sediment was fairly soft or not buried very deeply. Replacement origin can be shown by transaction of allochem structures by mosaic dolomite. Most definite replacement dolomite is coarser than .015 mm: it can be rather safely stated that if the massive dolomite is coarser than .03 mm it is certainly a replacement. Replacement of limestone may be complete or partial; in partially dolomitized limestones, the dolomite may occur as isolated grains, as winding tubules or irregular patches following zones of greater permeability.

Dolomites finer than .010 are believed to be of quasi-primary origin, accumulated directly as a dolomite ooze (dolomicrite). A directly precipitated origin is not proved, however, and it is possible that these beds were dolomitized very rapidly either at the sea bottom or during the first few cm of burial. These beds sometimes show small-scale cross-bedding and are frequently reworked into pebbles; this indicates that they were in existence as dolomite at a time when they were still close enough to the depositional interface to be eroded by stronger-than-usual waves.

Many factors control the precipitation of dolomite. Among those known to be important are temperature, solution, composition and concentration, rate of crystallization, and the presence and concentraton of certain organic compounds. But apparently the most important relations can be explained relatively simply by a diagram in which salinity is plotted against Mg/Ca ratio of the depositing solution. (Folk and Land '72 GSA abstra., 1975 AAPG).

The key idea is that dolomite is a very difficult mineral to form because of the precise Ca-Mg ordering required. Ideal dolomite consists of carbonate sheets interlayered with alternating sheets consisting entirely of Ca atoms or entirely of Mg atoms. These ions have such similar properties that it is difficult to segregate them, and segregation of ions to produce dolomite is accomplished more readily if the crystallization rate is very slow, or if solutions are dilute, with few interfering ions. Crystallization of dolomite is much more difficult if crystallization is rapid or if there is a high concentration of competing ions.

With a high concentration of Ca, Mg, and CO_3 ions, such as exists under evaporitic conditions, a highly oversaturated solution can lose energy by forming the improbable and difficult dolomite structure, or it can lose a smaller, but still substantial amount of energy by crystallizing as the simpler calcite or aragonite structures. In most cases the easier $CaCO_3$ mineral will form, and the Mg/Ca ratio in the remaining solution will rise to extreme values, as is observed in evaporitic environments. Only when extreme concentrations are reached is Mg finally forced to precipitate, as a poorly ordered, Ca-rich "protodolomite."

If the concentration of ions is much lower, such as in most meteoric waters, there is not nearly so much interference in crystal growth caused by lattice impurities. At slow rates of crystallization minerals can form that are more nearly in theoretical equilibrium with the surrounding solution; and in the case of dolomite the careful ordering of Ca and Mg which is required can be accomplished from solutions having Mg/Ca ratios near the theoretical value of 1:1.

In the diagram here presented the dolomite field is divided from the calcite field by a diagonal line indicating that, at low salinities (and low crystallization rates), dolomite can form readily at Mg/Ca ratios as low as 1:1, but as salinity and crystallization rate increase, the Mg/Ca ratio at which dolomite is first able to form rises until it must surpass 5:1 or 10:1 in hypersaline sabkhas. The line shows a generalized tendency and is a kinetic, not a thermodynamic, boundary; it is not meant to be a precise separator, because temperature, presence of various foreign ions, etc., can shift the precise position of this boundary.

<u>Hypersaline Environments</u>. Precipitation of gypsum and consequent loss of Ca raises the Mg/Ca ratio of the brine to values of 5:1 or more (by weight) sometimes as high as 100:1. Total salinity can rise from about five times that of normal sea water (at which point gypsum precipitates) to values as high as 10:1 or more. Kinsman (1965) has shown that hypersaline dolomite beings to crystallize only when the Mg/Ca ratio exceeds 5:1 to 10:1. Any carbonate crystallized below such high Mg/Ca ratios comes out as the easier-to-form aragonite or magnesian-calcite.

<u>Dolomite in Normally Saline Bays and Deep Sea</u>. Normal sea water has a Mg/Ca weight ratio of close to 3:1. Considerable controversy exists over whether dolomite can form in marine water of normal salinity, <u>e</u>. <u>g</u>., in nonrestricted bays or in deep-sea sediments. If dolomite can form in both hypersaline environments and in freshwater,

then it certainly must be able to form in sea water as well. Blatt et. al. asserted that theoretical calculations show that normal sea water should indeed be capable of producing dolomite, but that dolomite does not normally form because of the ordering difficulty.

Dolomite in Dilute Meteoric Water. Most freshwater lakes and streams average between 50-500 ppm total dissolved salts, whereas most large rivers contain 50-150 ppm. Most of these surface waters have Mg/Ca ratios between about 1:10 and 1:3; thus, calcite should be the expected phase formed. However, some exceptional lakes have an excess of Mg over Ca. Müller reported magnesian calcite, aragonite, and protodolomite forming in sediments of Lake Balaton, Hungary, with a total salinity under 500 ppm. But the Mg/Ca ratio in the lake varies from 1:1 to 3:1 and becomes even higher in the interstitial waters of the muds where protodolomite is present.

In caves, dolomite and other assorted magnesian carbonates can form if the Mg/Ca ratio exceeds 1:1. Dolomite can form in other freshwater deposits such as spring tufas.

In caliches calcite is again the normal product, as the Mg/Ca ratio of most soil waters is low. However, dolomitic caliche can form by attack of rain water upon Mg-rich source rocks.

We conclude that dolomite can form in surface fresh waters with Mg/Ca ratios as low as 1:1 providing crystallization rate is slow enough for ordering to take place. A more rapid crystallization rate requires a greater excess of Mg/Ca. A high total salinity is not in the least required; indeed, it tends to hamper crystallization of dolomite and favor calcite or aragonite instead.

Dolomite in Subsurface Waters. Subsurface waters show an extremely wide range in composition, from drinkable waters of as low as 100 ppm total salinity to very concentrated brines as high as six times seawater salinity (White, 1965). Because of a great loss of Mg relative to Ca in most sub-surface brines, the Mg/Ca ratio tends to be low, between 1:2 and 1:4.

Subsurface waters of mainly meteoric derivation, though dilute, have Mg/Ca ppm ratios that approach a limit very close to 1:2 (molar ratios of 0.8 Mg to 1.0 Ca). This is interpreted to be the point at which under sub-surface conditions calcite and dolomite are both stable, consequently precipitation of either phase can occur with slight shifts in composition. Of course, for large amounts of dolomite to be formed, enough time, adequate total supply of Mg, and actively moving waters must be available.

In summary, these data on natural waters show that, in a hypersaline environment with high ion concentration and rapid crystallization, the Mg/Ca ratio must exceed 5:1 or 10:1 for dolomite to form. In normal marine waters (such as deep-sea sediments), dolomite probably forms at Mg/Ca values over 3:1. In some fresh water and low-salinity subsurface waters, dolomite can form at Mg/Ca ratios as low as 1:1 because of a lack of competing ions and a generally slower rate of crystallization. The lower the salinity, the easier it is for dolomite to order.

On the diagram (fig. 1), this leads to a diagonal kinetic boundary line for the dolomite field. The fact that this boundary is diagonal is critically important. It means that the easiest way to produce dolomite is by lowering the salinity, reducing the concentration of competing ions, and in most cases slowing the rate of crystallization. Dolomite also can form at any given salinity by raising the Mg/Ca ratio, but that is not so common a process.

Methods for Producing Dolomite. The two most commonly accepted ways of forming dolomite are (1) with the hypersaline sabkha model, saline waters supplied either by supratidal flooding or by capillary upsucking; (2) by a reflux of heavy brines through porous sediments beneath the saline basins. Both these models rely upon using a brine that is always hypersaline. Hypersaline dolomite certainly exists, but we propose that it is much easier and of much greater importance to produce dolomite by dilution with fresh water. This can happen in several ways: (1) the schizohaline environment, wherein hypersaline brines are periodically mixed with fresh waters in a near-surface environment; (2) mixture in the shallow subsurface of evaporitic, highly saline brines with fresh water; (3) mixture of normal sea water or its connate equivalent with meteoric water, as in a salt-water/freshwater-lens contact zone; and (4) phreatic meteoric water collecting Mg purged from magnesian-calcite during diagenesis.

Schizohaline environment. Many shallow hypersaline environments are subjected to episodic flushing by fresh water. A sabkha can be flooded with monsoon rains; or a shallow hypersaline bay may be flushed by storms or hurricanes. In either case the salinity drops suddenly and drastically to nearly fresh conditions, only to be built slowly back to hypersaline conditions, as the normal evaporitic regime regains mastery. Such environments, characterized by wild swings between hypersaline and nearly fresh condition, with essentially no times of normal marine salinity, are designated as "schizohaline" environments.

Why is a schizohaline environment such an ideal place to form dolomite? Let us assume a typical sabkha with perhaps five times normal salinity and a Mg/Ca ratio of double normal, say 7:1. Adding fresh water drops the salinity drastically, but the Mg/Ca ratio remains almost as high as it was initially because of the low concentration of total salts, including Ca and Mg, carried by the diluting water. On the diagram this illustrated by a line dropping almost vertically and the compositon of the water plunges deeply into the dolomite field.

For example, diluting one part normal sea water with 9 parts of average river water with about 100 ppm total dissolved solids (typically including 20 ppm Ca and 7 ppm Mg) decreases the salinity of the mixture to about one-tenth-normal sea water, but changes the Mg/Ca ratio only from 3:1 to 2.2:1. Diluting a typical hypersaline sabkha brine with 9 parts of river water would drop the Mg/Ca ratio only from 7:1 to 6:1. Water of such composition in the subsurface then should continue to precipitate dolomite until the Mg/Ca ratio approaches 1:2.

Repetition of this process every several decades might allow significant quantities of dolomite to form. In such a dynamic environment of repeated flushing, effective transport of Mg ions through the system is insured. The large freshwater head developed during flooding may mix with and transport Mg-rich, diluted saline waters into the deep subsurface as a thick lens, and provide the "pump" necessary for massive diagenesis.

Subsidence or uplift--Even without the effect of flooding, gradual subsidence of hypersaline sabkha or lagoon sediments as deposition continues may allow penetration of meteoric waters from a landward freshwater table into the hypersaline-saturated sediments. A similar chemical revolution ensures. A fall in sea level will have the same effect, or uplift and emergency of sabkha sediments will raise the hypersaline system permanently into the vadose freshwater zone.

Mixing of fresh water with normal sea water--Many examples are known where a freshwater lens of huge size overlies sea water, with a zone of mixing in between characterized by dynamic water movement. In each case, mixing drastically lowers the salinity, but the Mg/Ca ratio remains high because the amounts of Ca and Mg added by fresh water are very small compared to the large amounts of these cations present in sea water. Mg/Ca ratios only can be lowered to typical subsurface values of 1:2 or 1:4 by crystallizing a large amount of dolomite or Mg-clays.

The failure to find gypsum in the supratidal flats of the Bahamas suggests that some of the dolomite crusts may be formed by wet-season rain-water dilution of sea water that is either of normal salinity or that has not yet evaporated down to the gypsum stage; thus, the cause of the dolomite is, again, freshwater dilution.

Mg-purging--Incongruent dissolution of Mg calcites can cause dolomitization at nearly constant salinity by increasing the Mg/Ca ratio of the pore waters (Land and Epstein, 1970). Only minor amounts of dolomite, restricted to Mg-calcite allochems or their internal pores, can be produced by this method, however.

Limpid Dolomite: Produce of Dilute Waters. Freshwater dolomite is characteristically limpid (Folk and Land, 1972; Siedlecka, 1972; Land, 1973). Crystals are readily identifiable with a low-power binocular microscope (or even hand lens or naked eye) because of their mirror-smooth faces, reflecting sunlight like tiny faceted gems. Crystals are usually perfectly brillant, transparent, and water-clear, lacking any inclusions; they look like minute diamonds. Under the petrographic microscope they appear as perfect euhedra with geometrically exact faces, and again are extremely clear though some are zoned. At higher magnifications obtained by electron microscope they appear as near-perfect crystals with absolutely plane crystal faces, lacking any imperfections, inclusions, growth steps, or any other sort of blemish.

Conclusions. Dolomite, because of the difficulty of ordering required for crystallization, can form most easily by slow crystallization. At more rapid crystallization rates, aragonite and magnesian calcite crystallize because they are simpler structures. Dolomite is favored similarly when solutions are dilute because few impurities disrupt the precise ordering of the lattice.

Dolomite can form from any natural solution providing the Mg/Ca ratio is over approximately 1:1, even in lakes or subsurface waters of very low total salinity. An important way to precipitate dolomite is to dilute sea water or sabkha-evaporitic water with fresh water. Dilution allows the Mg/Ca ratio to remain very high, but slows the crystallization rate and reduces the concentration of competing ions. Two ideal sites where such a mixing mechanism can take place (schizohaline environments) are floodable sabkhas or inundatable shallow lagoons where salinity undergoes rapid fluctuation between hypersaline and nearly fresh conditions; another site is the subsurface zone where sea water or evaporitic waters come into contact with a wedge or lens of meteoric water and salinity reduction occurs. In both cases Mg is supplied by saline waters, but precipitation is permitted only by dilution with fresh waters.

SPHERICITY-FORM DIAGRAM FOR PARTICLE SHAPES

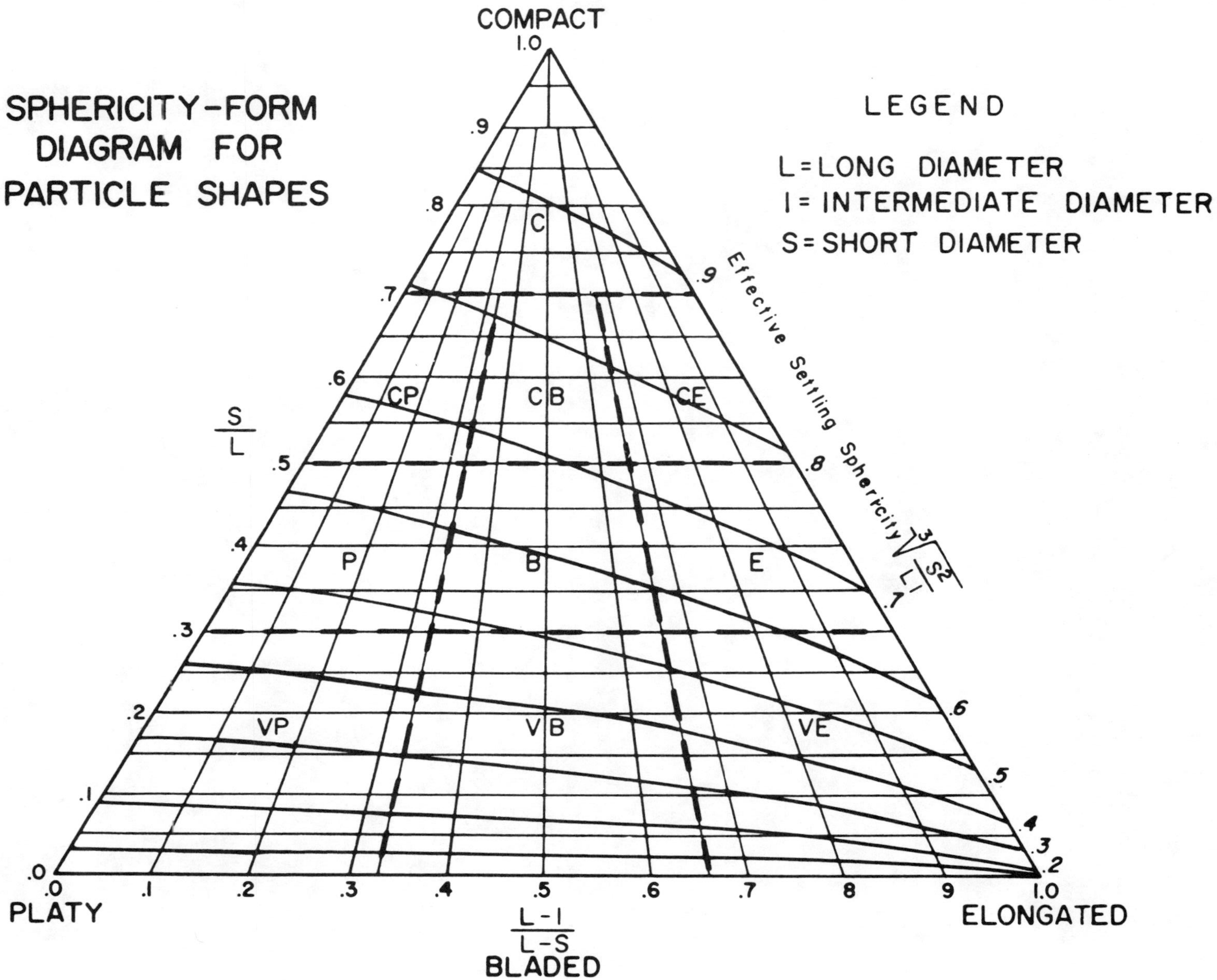

Roundness Scale of Maurice Powers (1953) - J. Sed. Petr. 23:117-119

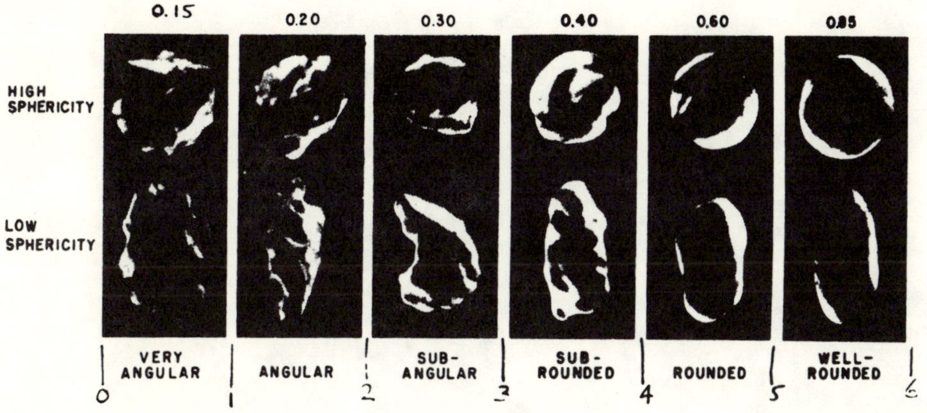

Logarithmic Transform - ρ Scale of Folk - J. Sed. Petr. 25:297-301

Fig. 1 - Roundness scale. (Powers)